*Oct-1980.*

# Physics Programs

---

*Magnetism*

---

# Physics Programs

*Magnetism*

*Edited by*

**A. D. Boardman**

*Department of Pure and Applied Physics, University of Salford*

*A Wiley–Interscience Publication*

## JOHN WILEY & SONS

Chichester · New York · Brisbane · Toronto

*British Library Cataloguing in Publication Data:*

Physics programs.
  Magnetism
  1. Physics—Programmed instruction
  I. Boardman, A. D.
  530′.07′7    QC21.2    80–40124

  ISBN 0 471 27733 9


Typeset in Northern Ireland, at The Universities Press (Belfast) Ltd. and printed by Pitman Press, Bath

# Contents

# Contents

# Preface

This small book is the magnetism section of the larger textbook called *Physics Programs* that covers the four areas, optics, magnetism, solid state/quantum physics, and applied physics. Each chapter given here is self-contained, with enough theory given for the topic discussed, and the associated computer programs, to be well understood. The programs are guaranteed in the sense that they are copied directly from fully working source texts on the computer. They can be used, possibly with minor adjustments, on any computing system. If what is required is a classroom demonstration, or the engagement of a class in a simple sequence of exercises, then the programs may be used without understanding the coding. The programs are, however, liberally strewn with comments so that they can be used for more advanced projects in which an understanding of the program is required.

The material included here deals with three widely differing topics in magnetism. The first chapter discusses the classical computation of the magnetostatic field near a permanent magnet while the second chapter discusses high gradient magnetic separation, a topic of some practical importance. The final chapter deals with a topic from advanced solid state physics and is a study of magnetization in the rare earth, praseodymium. It is hoped that this set of chapters will appeal to all teachers and undergraduate and graduate students involved in magnetism courses.

*Salford*                                                      A. D. BOARDMAN

Physics Programs
Edited by A.D. Boardman
© 1980 John Wiley & Sons Ltd.

CHAPTER 4

# Calculation of the Fields Near Permanent Magnets

M. I. DARBY

## 1. INTRODUCTION

Partial differential equations are important in almost all branches of physics, and often they can only be solved numerically. Owing to the diversity of boundary conditions and other factors that may apply, it is impracticable to produce computer library routines capable of solving more than one specific type of problem. For this reason it is valuable to have some practical experience of the difficulties involved in applying one of the common numerical techniques in a relatively simple situation.

The problem considered here is the calculation of the magnetic field in the vicinity of a uniformly magnetized rectangular permanent magnet. The magnet is assumed to be infinite in one direction, so that the problem reduces to two dimensions. The basic magnetostatic equations are given in sections 2 and 3. There it is shown that the fields are conveniently written in terms of a scalar magnetostatic potential, which satisfies Laplace's equation, and which is completely determined by the boundary conditions. Sometimes it is possible to obtain an analytical solution for the potential, but usually Laplace's equation must be solved numerically. One method of doing so, and that adopted here, is to replace the partial differential equation with a set of (linear) finite difference equations. These can then be solved by standard methods, either directly by elimination or by iteration. The latter method is employed below.

## 2. THE MAGNETOSTATIC POTENTIAL

The magnetic induction vector $\mathbf{B}$ produced by a steady electric current $I$ satisfies Ampère's law,[1]

$$\oint_C \mathbf{B} \cdot d\mathbf{l} = \mu_0 I, \tag{1}$$

where $C$ is a contour enclosing the conductor carrying $I$. Employing Stokes's integral theorem, this equation can be written as

$$\text{curl } \mathbf{B} = \mu_0 \mathbf{J}, \tag{2}$$

where $\mathbf{J}$ is the current density $(\text{Am}^{-2})$. The other basic property of $\mathbf{B}$ is that it forms closed loops, i.e. it satisfies

$$\text{div } \mathbf{B} = 0. \tag{3}$$

A small current loop produces a field $\mathbf{B}$ which resembles the electric field near an electric dipole, and consequently a magnetic dipole moment can be identified with the loop. A magnetic material may be thought of as containing a large number of elementary loops, giving rise to a dipole moment per unit volume, $\mathbf{M}$, known as the magnetization. The magnetization contributes to $\mathbf{B}$ and it can be shown[1] quite generally that equation (2) is replaced by

$$\text{curl } \mathbf{B} = \mu_0 \mathbf{J} + \mu_0 \text{ curl } \mathbf{M}, \tag{4}$$

where $\mathbf{J}$ is the real current density. It is convenient to define a magnetic field $\mathbf{H}$ by

$$\mathbf{B} = \mu_0(\mathbf{H} + \mathbf{M}), \tag{5}$$

and from equation (4) $\mathbf{H}$ satisfies

$$\text{curl } \mathbf{H} = \mathbf{J}, \tag{6}$$

There are usually no true currents in a permanent magnet so that (6) reduces to

$$\text{curl } \mathbf{H} = 0, \tag{7}$$

and therefore it is possible to define a scalar magnetic potential $\phi$ by

$$\mathbf{H} = -\nabla\phi. \tag{8}$$

From equations (3) and (5),

$$\text{div } \mathbf{H} = -\text{div } \mathbf{M}, \tag{9}$$

or in terms of $\phi$,

$$\nabla^2\phi = \text{div } \mathbf{M}. \tag{10}$$

This is Poisson's equation for the potential and, by analogy with electrostatics, the term div $\mathbf{M}$ plays the role of a volume magnetic charge density, and is frequently referred to as the pole density.

## 3. BOUNDARY CONDITIONS ON INTERFACES

The boundary conditions on the magnetostatic potential at the interface between two media in which true currents are absent can be derived from

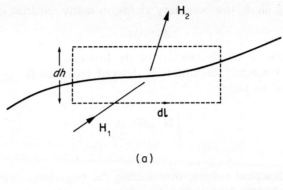

(a)

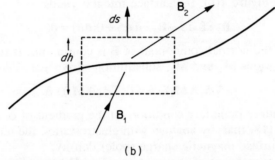

(b)

Figure 1. Fields near the boundary between two
media illustrating the geometry and notation employed in the text to discuss the boundary conditions
on (a) **H**, and (b) **B**

equations (7) and (3). Integration of (7) yields

$$\oint \mathbf{H} \cdot d\mathbf{l} = 0, \tag{11}$$

and evaluating the line integral around a rectangular contour intersecting
the boundary between two regions, in which the magnetic fields are $\mathbf{H}_1$ and
$\mathbf{H}_2$, as shown in Figure 1(a),

$$\mathbf{H}_1 \cdot d\mathbf{l} - \mathbf{H}_2 \cdot d\mathbf{l} + 0(dh) = 0, \tag{12}$$

(assuming that $dh$ can be made arbitrarily small). Equation (12) implies that
the tangential components of **H** are continuous across the boundary, and can
also be expressed as

$$\hat{\mathbf{n}} \times (\mathbf{H}_1 - \mathbf{H}_2) = 0, \tag{13}$$

where $\hat{\mathbf{n}}$ is a unit vector normal to the surface. In terms of the scalar
potentials,

$$\hat{\mathbf{n}} \times (\nabla \phi_1 - \nabla \phi_2) = 0 \tag{14}$$

and, integrating along the boundary yields, in many circumstances,

$$\phi_1 = \phi_2. \tag{15}$$

Thus the potential is continuous across the boundary.

The second boundary condition is derived from $\operatorname{div} \mathbf{B} = 0$ by applying Gauss's theorem to yield

$$\int_S \mathbf{B} \cdot d\mathbf{S} = 0. \tag{16}$$

For a small cylindrical volume intersecting the boundary, indicated by the dotted lines in Figure 1(b), the surface integral yields

$$\mathbf{B}_2 \cdot \hat{\mathbf{n}} \, dS - \mathbf{B}_1 \cdot \hat{\mathbf{n}} \, dS + 0(dh) = 0, \tag{17}$$

indicating that the normal component of $\mathbf{B}$ is continuous. If the two regions have magnetizations $\mathbf{M}_1$ and $\mathbf{M}_2$, substitution of $\mathbf{B} = \mu_0(-\nabla\phi + \mathbf{M})$ yields

$$(-\nabla\phi_1 + \mathbf{M}_1) \cdot \hat{\mathbf{n}} = (-\nabla\phi_2 + \mathbf{M}_2) \cdot \hat{\mathbf{n}}. \tag{18}$$

This is the required boundary condition on the gradient of $\phi$. It can be seen from equation (18) that, by analogy with electrostatics, the term $\mathbf{M} \cdot \hat{\mathbf{n}}$ plays the role of a surface magnetic charge (pole) density.

The normal component of the magnetic field $\mathbf{H}$ has a discontinuity equal to the difference of the components of the magnetizations. Consequently the magnetic field inside the magnet opposes the magnetization and is known as the demagnetizing field.

## 4. THE MODEL PROBLEM

The computational problem is to determine the magnetic field in the regions inside and outside a two-dimensional rectangular magnet by solving Poisson's equation (10) for the scalar potential. It is assumed that the magnet is uniformly magnetized, so that $\mathbf{M}(\mathbf{r})$ is a constant vector, in which case equation (10) reduces to Laplace's equation,

$$\frac{\partial^2 \phi}{\partial x^2} + \frac{\partial^2 \phi}{\partial y^2} = 0, \tag{19}$$

everywhere except on the boundary of the magnet. On the latter, the potential is continuous, but the components of its gradient normal to the surface change by the normal component of the magnetization, in accordance with equation (18).

At large distances from the magnet the potential will resemble that of a small magnetic dipole of moment $\mathbf{m} = \mathbf{M}V$, where $V$ is the volume of the

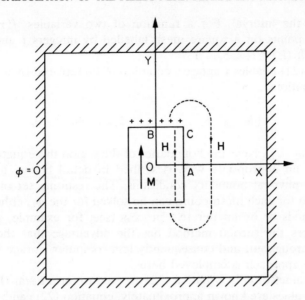

Figure 2. Schematic diagram showing the positions of the magnet and the outer boundary, and indicating the directions of the magnetic field. Because of symmetry it is necessary to consider only the region in the positive quadrant

magnet, i.e.

$$\phi(\mathbf{r}) = -\frac{1}{4\pi}\,\mathbf{m}\cdot\mathrm{grad}\left(\frac{1}{r}\right). \tag{20}$$

In principle it is possible to match the numerical solutions of (19) to this expression on a distant rectangular boundary, but to avoid this added complication it will be assumed here that $\phi$ is essentially zero on that boundary, as shown in Figure 2. The effect of this approximation on the final solutions can be investigated by increasing the size of the large rectangle.

## 5. THE FINITE DIFFERENCE EQUATIONS

Laplace's equation (19) can be solved by approximating the derivatives of $\phi$ by finite difference formulae. The second derivative of a function of a single variable $f(x)$ tabulated at equal intervals of $x$ can be approximated,[2] using Taylor's expansion, by

$$\frac{\mathrm{d}^2 f}{\mathrm{d}x^2} \approx \frac{1}{h^2}\{f(x+h)+f(x-h)-2f(x)\}, \tag{21}$$

where $h$ is the interval. For a function of two variables, $f(x, y)$ can be specified at points on a square mesh labelled by integers $i$ and $j$, so that $x = ih$; $y = jh$ $(i, j = 1, 2, 3, \ldots)$.

Equation (21) enables Laplace's equation to be replaced by a set of finite element equations:

$$\frac{1}{h^2} \{\phi_{i+1,j} + \phi_{i-1,j} + \phi_{i,j+1} + \phi_{i,j-1} - 4\phi_{i,j}\} = 0, \tag{22}$$

for each point $(i, j)$. Near the boundaries of the region this equation must be modified in an appropriate way, described in detail below, to take into account the physical boundary conditions. The resulting set of equations, one equation for each mesh point, can be solved for the $\phi_{i,j}$ either by direct matrix methods or by an iterative process (see, for example, ref. 3). For large matrices the second method has the advantage that the zeros are preserved throughout, and consequently less computer storage is required. An iterative approach is employed here.

The five values of $\phi$ in equation (22) are said to form a star (Figure 3). If four of the values are known approximately, equation (22) can be employed to determine an improved value for the fifth. In an iterative process initial values of $\phi$, $\phi_{i,j}^{(0)}$ say, must be assigned to each mesh point. Generally, it is not essential that these initial values are a close approximation to the final solution. On the boundaries the $\phi$-values may be known exactly from the outset, but often the $\phi_{i,j}^{(0)}$-values inside the region can only be chosen somewhat arbitrarily. Frequently the $\phi_{i,j}^{(0)}$ are set equal to a constant value

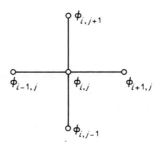

Figure 3. The star of function values required to approximate $\nabla^2 \phi$ at a general mesh point $(i, j)$. The mesh is assumed to be square and the mesh size is $h$

(0.5 in the program presented here). The initial $\phi_{i,j}^{(0)}$ can be improved by applying equation (22) to each mesh point in turn, giving quantities $\phi_{i,j}^{(1)}$. Explicitly,

$$\phi_{i,j}^{(1)} = \tfrac{1}{4}\{\phi_{i+1,j}^{(0)} + \phi_{i-1,j}^{(0)} + \phi_{i,j-1}^{(0)} + \phi_{i,j+1}^{(0)}\}, \qquad (23)$$

If the new $\phi_{i,j}^{(1)}$ are substituted into equation (22) the bracket on the left-hand side will not be exactly zero, but will have a residual value, $R_{ij}$ say, which is some measure of the discrepancy between $\phi_{i,j}^{(1)}$ and the true solution $\phi$. Repeating the procedure, new values $\phi_{i,j}^{(2)}$ can be computed from the $\phi_{i,j}^{(1)}$ using a formula similar to (23). This iterative process is continued until the $\phi$ values do not alter, within a specified accuracy, between one scan of the mesh points and the next.

Iterative formulae like (23) employing only the old $\phi^{(0)}$ values on the right-hand side are said to be of Jacobi type. When performing the calculations with a computer it is more natural to use the newly calculated $\phi$ values in the right-hand side of (23) as soon as possible. Thus, for example, if the mesh is scanned column by column the quantities $\phi_{i-1,j}^{(1)}$ and $\phi_{i,j-1}^{(1)}$ will have been computed before $\phi_{i,j}^{(1)}$ is evaluated, and equation (23) can be replaced by

$$\phi_{i,j}^{(1)} = \tfrac{1}{4}\{\phi_{i+1,j}^{(0)} + \phi_{i-1,j}^{(1)} + \phi_{i,j-1}^{(1)} + \phi_{i,j+1}^{(0)}\}. \qquad (24)$$

This expression gives rise to a Gauss–Seidel scheme. It can be shown[3] that this iterative process converges more rapidly than the simpler Jacobi method.

More complicated iteration formulae than (24) have been devised which give even more rapid convergence than the Gauss–Seidel method. One such procedure,[3] which changes the old field value by adding to it a small fraction of the old residual $R_{ij}$ is known (amongst other names) as successive over-relaxation (SOR). For the $n$th iteration, assuming column-by-column scanning of the mesh, the appropriate formula is

$$\phi_{i,j}^{(n)} = \phi_{i,j}^{(n-1)} + \frac{\alpha}{4}\{\phi_{i+1,j}^{(n-1)} + \phi_{i-1,j}^{(n)} + \phi_{i,j-1}^{(n)} + \phi_{i,j+1}^{(n-1)} - 4\phi_{i,j}^{(n-1)}\}, \qquad (25)$$

where $\alpha$ is a parameter which usually lies between 1 and 2 in practice.[3] This method has been employed in the present work.

Because of the symmetry of the problem it is necessary to consider only one-quarter of the total region, for example the first quadrant shown in Figure 4. The iteration formulae (23)–(25) cannot be employed for points on the surface of the magnet because Laplace's equation is not valid there. Nor can these formulae be used as they stand for points on the boundaries of the region, because some of the $\phi$-values in the star formulae for the boundary points will be outside the region. The latter situation can be illustrated by

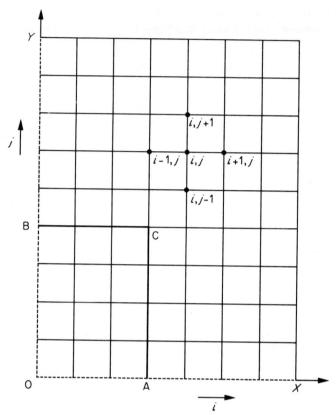

Figure 4. A typical mesh covering the region of interest. The area $OACB$ represents the portion of the magnet in this quadrant

considering the boundary $OY$, for which $i = 1$, giving for the residuals $R_{1j}$

$$R_{1j} = \phi_{2,j} + \phi_{0,j} + \phi_{1,j+1} + \phi_{1,j-1} - 4\phi_{1,j}. \qquad (26)$$

The values $\phi_{0,j}$ lying outside the field region are sometimes called 'fictitious' values, and are often labelled by an asterisk added as a superscript, e.g. $\phi_{0,j}^{*}$. Because of the special conditions applying at the boundaries the fictitious values required to calculate the residuals can usually be expressed as functions of the $\phi_{i,j}$ values inside the region. For the present problem the appropriate special finite difference formulae required for the region boundaries and the magnet surface are now derived in detail.

## 5.1 The outer boundary

All $\phi_{i,j} = 0$, and this boundary is excluded from the iterative process.

## 5.2 The symmetry axis $OY$

The line $OY$ lies along the $y$-coordinate axis (Figure 4). The field in the negative $x$ region $(y > 0)$ is the mirror image of that in the first quadrant, i.e.

$$\phi(-x, y) = \phi(x, y).$$

Hence the fictitious field values $\phi_{0,j}^*$ (see Figure 5a) are given by

$$\phi_{0,j}^* = \phi_{2,j}. \tag{27}$$

Employing this in the star formula for $R_{ij}$ gives a SOR formula:

$$\phi_{1,j}^{(n)} = \phi_{1,j}^{(n-1)} + \frac{\alpha}{4}\{2\phi_{2,j}^{(n')} + \phi_{1,j-1}^{(n')} + \phi_{1,j+1}^{(n')} - 4\phi_{1,j}^{(n-1)}\}, \tag{28}$$

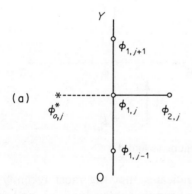

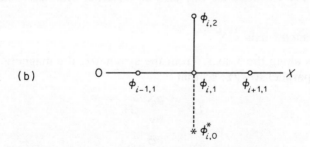

Figure 5. Function values involved in treating typical points on the boundaries: (a) symmetry line $OY$ showing the position of the fictitious point $\phi_{0,j}^*$; (b) symmetry line $OX$ showing the position of the fictitious point $\phi_{i,0}^*$; (c) magnet boundary $AC$; (d) magnet boundary $BC$

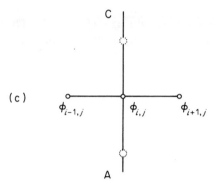

(c)

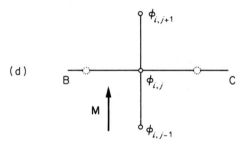

(d)

Figure 5c and d

where $n'$ is $n$ or $(n-1)$ and indicates that the most recently calculated $\phi$-values are to be inserted. This formula applies for $j > 1$, and the origin $O$ and point $B$ are excluded as they also lie on other symmetry axes.

### 5.3 The symmetry axis $OX$

Line $OX$ lies along the $x$-axis. From the symmetry, the magnetic field $\mathbf{H}$ is everywhere parallel to $OY$, so that

$$H^y = -\frac{\partial \phi}{\partial y} = H, \tag{29}$$

$$H^x = -\frac{\partial \phi}{\partial x} = 0, \tag{30}$$

The second equation means that all the $\phi$-values on $OX$ are equal. Furthermore, the continuity of $\mathbf{H}$ across $OX$ requires that the gradient $\partial \phi/\partial y$ should be continuous. Defining fictitious values $\phi_{i,0}^*$ as in Figure 5(b), and employing a simple forward difference formula for the gradient, this condition gives

$$\frac{1}{h}\{\phi_{i,2}-\phi_{i,1}\}=\frac{1}{h}\{\phi_{i,1}-\phi_{i,0}^*\},$$

or

$$\phi_{i,0}^*=2\phi_{i,1}-\phi_{i,2}. \tag{31}$$

Substitution into the star formula for the residual, the SOR iteration formula becomes

$$\phi_{i,1}^{(n)}=\phi_{i,1}^{(n-1)}+\frac{\alpha}{4}\{\phi_{i+1,1}^{(n')}+\phi_{i-1,1}^{(n')}-2\phi_{i,1}^{(n-1)}\}, \tag{32}$$

where $n'$ is $n$ or $(n-1)$ depending on how the boundary mesh points are scanned. In fact this equation is redundant because the $\phi$ are all equal by equation (30). Hence, since $\phi$ is zero on the outer boundary, it is necessary to set

$$\phi=0,$$

at all points on the line, including the origin $O$.

## 5.4  The magnet boundary $AC$

The condition on tangential $\mathbf{H}$ requires that $\phi$ is continuous on the boundary, and this is automatically ensured in the iteration. Since the magnetization has no component perpendicular to $AC$, the continuity of the normal component of $\mathbf{B}$ implies that $\partial\phi/\partial x$ is continuous. Employing a forward difference formula, see Figure 5c,

$$\frac{1}{h}\{\phi_{i+1,j}-\phi_{i,j}\}=\frac{1}{h}\{\phi_{i,j}-\phi_{i-1,j}\},$$

or

$$\phi_{i,j}=\tfrac{1}{2}\{\phi_{i+1,j}+\phi_{i-1,j}\}. \tag{33}$$

Hence on this boundary, excluding point $C$, which requires a different treatment because it also lies on boundary $BC$,

$$\phi_{i,j}^{(n)}=\phi_{i,j}^{(n-1)}+\frac{\alpha}{2}\{\phi_{i+1,j}^{(n')}+\phi_{i-1,j}^{(n')}-2\phi_{i,j}^{(n-1)}\}, \tag{34}$$

where again $n'$ is either $n$ or $(n-1)$ as appropriate.

## 5.5 The magnet boundary $BC$

Again $\phi$ is continuous, but now, from equation (18), the gradient in y has a discontinuity equal to $|\mathbf{M}|$, i.e.

$$-\frac{\partial\phi}{\partial y}\bigg|_{in} + M = -\frac{\partial\phi}{\partial y}\bigg|_{out}. \tag{35}$$

In terms of finite differences, this becomes (see Figure 5d):

$$-\frac{1}{h}\{\phi_{i,j+1} - \phi_{i,j}\} = -\frac{1}{h}\{\phi_{i,j} - \phi_{i,j-1}\} + M,$$

or

$$\phi_{i,j} = \tfrac{1}{2}\{\phi_{i,j+1} + \phi_{i,j-1} + Mh\}. \tag{36}$$

Therefore, excluding point $C$, but including point $B$,

$$\phi_{i,j}^{(n)} = \phi_{i,j}^{(n-1)} + \frac{\alpha}{2}\{\phi_{i,j+1}^{(n')} + \phi_{i,j-1}^{(n')} + Mh - 2\phi_{i,j}^{(n-1)}\}. \tag{37}$$

## 5.6 The point $C$

This point is difficult to treat satisfactorily. Only a very approximate expression for $\phi$ will be used here, it being assumed that a simple average of the expressions for $AC$ and $BC$, given by (33) and (36), is appropriate. Hence, adding,

$$2\phi_{i,j} = \tfrac{1}{2}\{\phi_{i+1,j} + \phi_{i-1,j}\} + \tfrac{1}{2}\{\phi_{i,j+1} + \phi_{i,j-1} + Mh\}, \tag{38}$$

and

$$\phi_{i,j}^{(n)} = \phi_{i,j}^{(n-1)} + \frac{\alpha}{4}\{\phi_{i+1,j}^{(n')} + \phi_{i,j+1}^{(n')} + \phi_{i,j+1)}^{(n')} + \phi_{i,j-1}^{(n')} + Mh - 4\phi_{i,j}^{(n-1)}\}. \tag{39}$$

## 6. THE COMPUTER PROGRAM

The iteration scheme is readily programmed for a computer, and a suitable FORTRAN coding is given at the end of the chapter. The flow diagram Figure 6 indicates a few practical details. The iteration loop for the $\phi_{i,j}$ is included in the main routine. Subroutines calculate the magnetic field and handle the output of data.

It is necessary to keep a running count of the number of iterations performed, so that the process can be terminated if it shows no sign of converging to the required accuracy. There are several criteria for satisfactory convergence that can be employed, and in the appended program two different criteria are used in series. In the first, the residual with the largest

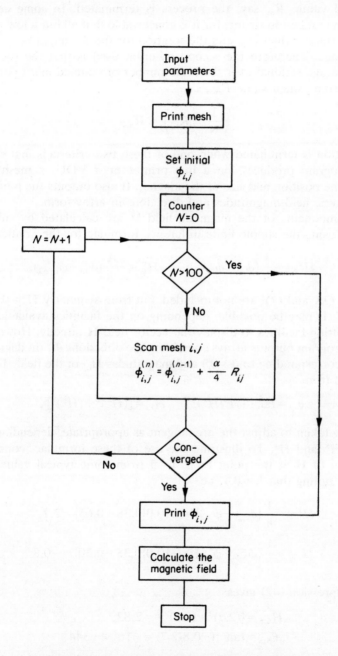

Figure 6. Flow diagram for the computer program

magnitude is determined for the $n$th iteration, and if this is less than some prescribed value, $R_m$ say, the process is terminated. In some cases this criterion might be too strong, for it is conceivable that all but a few residuals are much smaller than $R_m$, and the solution for the $\phi_{i,j}$ might be acceptable at that stage. Therefore the second criterion used is that the root-mean-square average residual over the total number of scanned mesh points $N$ is less than some small value $R_{ms}$, i.e.

$$\left(\frac{1}{N}\sum_{i,j} R_{i,j}^2\right)^{\frac{1}{2}} < R_{ms}. \tag{40}$$

The iteration is terminated when one of these two criteria is first satisfied.

The program produces, on a line printer or a VDU, a mesh pattern showing the position and size of the magnet. It also outputs the potential $\phi_{ij}$ and magnetic field magnitudes and directions in array form.

The components of the magnetic field $\mathbf{H}$ are calculated by subroutine THEMA using the simple finite difference formulae for the gradients:

$$H_{i,j}^x = -\frac{1}{h}(\phi_{i,j} - \phi_{i-1,j}); \qquad H_{i,j}^y = -\frac{1}{h}(\phi_{i,j} - \phi_{i,j-1}). \tag{41}$$

The lines $OX$ and $OY$ are not included, but from symmetry $H^x = 0$ in these directions. It may be possible, depending on the facilities available, to use graph plotting facilities to display the vector field $\mathbf{H}$ directly. However, the present program outputs arrays containing the directions $\theta_{i,j}$ (in degrees with line $OX$ corresponding to $\theta = 0°$) and magnitudes $H_{i,j}$ of the field. These are calculated from

$$\theta_{i,j} = \tan^{-1}(H^y/H^x); \qquad H_{i,j} = [(H^x)^2 + (H^y)^2]^{\frac{1}{2}}, \tag{42}$$

care being taken to adjust the arc tangent as appropriate, depending on the signs of $H^x$ and $H^y$. To illustrate the use of these formulae, consider the calculation of $\mathbf{H}$ at the point $i = 5$, $j = 5$ from some typical values of the potential, noting that $h = 0.1$, i.e.

$$H^x = -\frac{1}{h}\{\phi_{5,5} - \phi_{4,5}\} = -10.0\{0.38 - 0.65\} = 2.7,$$

$$H^y = -\frac{1}{h}\{\phi_{5,5} - \phi_{5,4}\} = -10.0\{0.38 - 0.30\} = -0.8. \tag{43}$$

Hence expression (42) gives

$$H_{5,5} = \{(2.7)^2 + (0.8)^2\}^{\frac{1}{2}} = 2.82,$$

$$\theta_{5,5} = \tan^{-1}(-0.8/2.7) = -16° \equiv +344°. \tag{44}$$

These answers can be located in the typical output given in this chapter.

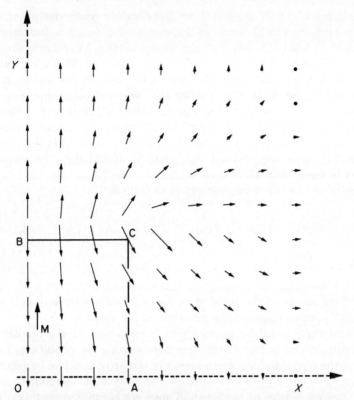

Figure 7. A typical magnetic field distribution in the positive quadrant plotted from results obtained from the computer program

The values of $H_{i,j}$ and $\theta_{i,j}$ enable the vector magnetic field to be plotted at each mesh point. For the mesh point $(5, 5)$ for example, the vector $\mathbf{H}$ can be represented by a line of length 2.82 units (on some suitable scale) drawn to make an angle of $+344°$ with the $OX$-axis. A plot of all the field vectors corresponding to the typical output is shown in Figure 7.

## 7. RUNNING THE PROGRAM

The small list of input parameters is as follows:

IN, JN    Number of mesh points in the $x$- and $y$-directions, respectively.
IM, JM    Number of mesh points in the $x$- and $y$-directions occupied by the magnet.
H         Mesh size (metres) (written as $h$ in the text).
ALPHA     Convergence parameter $(1 < \alpha < 2)$.
AMAG      The magnetization $(\text{Am}^{-1})$.

The basic equations of section 2 are appropriate when the physical quantities are measured in SI units. In this system the magnetic induction vector **B** has units of tesla (T), and both the magnetization **M** and magnetic field **H** have units of ampere metre$^{-1}$ (Am$^{-1}$). The values of $|\mathbf{M}|$ for typical permanent magnets are in the range $10^5$–$10^6$ Am$^{-1}$. Spatial dimensions are measured in metres. Since $M$ and $h$ enter the potential equations only through the product $Mh$, the solution for a given value of $Mh$ is applicable to many problems with different combinations of these parameters. However, it should be noted that the magnitude of the magnetic field **H** depends on the value of $h$ in each case, through the gradient, and further, the magnitude of **H** scales linearly with $M$.

A typical set of input parameters is as follows:

| 9 | 10 | 4 | 5 | 0.1 | 1.5 | 5.0 |
|------|------|------|------|------|---------|--------|
| (IN) | (JN) | (IM) | (JM) | (H) | (ALPHA) | (AMAG) |

The first six parameters are read in on one line using FORMAT (4I4, 2F5.2) and AMAG is read separately using FORMAT (F5.2). It should be noted that the same device number has been employed for both input and output. The values of the input data are printed out immediately followed by a mesh pattern showing the position of the magnet.

The next three numbers printed give information on the convergence of the iteration process. The parameters determining the conditions for termination of the iteration have been preset in the program. The largest residual, REMAX, and the root mean square of the residuals, RESUM, are determined after each scan of the mesh. These are compared with two parameters, EPMAX and EPSUM respectively, which have both been given values of $10^{-4}$. If REMAX is less than EPMAX, or if RESUM is less than EPSUM, it is assumed that the solution is sufficiently accurate for the present purpose. From the typical results given here it is seen that iteration has terminated when RESUM ($=0.95 \times 10^{-4}$) became smaller than $10^{-4}$. The largest residual at that time was reasonably close to $10^{-4}$ (REMAX $= 2.4 \times 10^{-4}$), giving confidence in the solution obtained. The number of iterations needed is also printed out, being 28 in the example given. A control in the program stops the calculations if the integer NCOUN, counting the number of iterations, exceeds 100. It is a simple matter to change the preset parameters by making minor modifications to the program coding.

Finally, the arrays containing $\phi_{i,j}$, $\theta_{i,j}$ and $H_{i,j}$ are printed out. To understand the solution for the vector field **H**, the latter should be plotted in a form like Figure 7. It can be seen from that figure that inside the magnet the field is almost parallel to the direction of magnetization, but is in the opposite direction. This is the demagnetizing field. Outside the magnet the field is strongest near the magnet's face $BC$, where it is roughly parallel to the direction of **M**. This behaviour is that expected from the simplest model of a bar magnet in which (fictitious) magnetic poles are placed on the faces. The field falls off rapidly with distance in the $x$-direction, and smoothly changes direction, except near the corner of the magnet $C$, until on the symmetry axis $OX$ it is in the opposite direction to **M**.

There are several exercises that can be performed with the computer program given here, simply by changing the input parameters. For instance it is instructive to investigate how the magnetic field distribution depends on the various constants, and how the convergence of the iterative process depends on $\alpha$. Four typical exercises of this type are:

(1) Determination of the dependence of the magnetic field distribution on the size, shape, and magnetization of the magnet (vary input parameters IN, IM, JN, JM and AMAG).

(2) Investigation of the effects of changing the mesh size $h$ and the determination of the optimum size for $h$ (vary H in conjunction with IN, IM, JN and JM to retain the same region and magnet sizes).

(3) Investigation of the sensitivity of the magnetic field distribution near the magnet to the distance of the outer boundary from the magnet (vary IN and IM). Ideally the far boundary should be at infinity.

(4) Determination of the dependence of the rate of convergence of the iterative process for the $\phi_{i,j}$ on the value of the successive over-relaxation parameter (vary ALPHA) (cf. ref. 3 for the theory).

More insight into the method of solution of the problem can be obtained by making modifications to the program. Five typical exercises, involving only minor changes to the coding, could be based on using the following:

(a) other convergence parameters EPMAX and EPSUM;

(b) changing the coding to calculate the magnetic induction field **B** (from equation (5));

(c) more complicated difference formulae;[4]

(d) better formulae for special points,[4] e.g. for point $C$;

(e) other convergence criteria.[3,4]

The next stage of investigation would involve considering different models requiring considerable changes to the basic formulae, and effectively requiring completely new programs to be written. Exercises of this type are more advanced and could form the basis of project work. Typical examples include the consideration of the following:

(i)   magnets of different shapes (but with plane boundaries);

(ii)  composite rectangular magnets of materials with different magnetizations;

(iii) the effect of using a different mesh size over some part of the region of the field;

(iv)  two or more separated rectangular magnets;

(v)   cylindrical magnets.

Having moved away from finite difference formulae in Cartesian coordinates, as in (v) above, many other projects are possible. There are also problems in many other areas of physics that require solutions to elliptical partial differential equations obtainable by a similar numerical treatment to that considered here.

## REFERENCES

1. W. K. H. Panofsky and M. Philips, *Classical Electricity and Magnetism* (Addison-Wesley, Reading, Mass., 1962).
2. C. D. Smith, *Numerical Solution of Partial Differential Equations* (Oxford University Press, London, 1965).
3. *Modern Computing Methods*, 2nd edn. (H.M.S.O., London, 1961).
4. K. J. Binns and P. J. Lawrenson, *Analysis and Computation of Electric and Magnetic Field Problems* (Pergamon Press, London, 1963).

PERMANENT MAGNETS

```
      DIMENSION PHI(50,50),TH(50,50),FE(50,50)
C   INPUT AND PRINT OUT PARAMETERS
      READ(1,100)IN,JN,IM,JM,H,ALPHA
      READ(1,101)AMAG
  100 FORMAT(4I4,2F5.2)
  101 FORMAT(F5.2)
      WRITE(1,105)IN,JN
      WRITE(1,106)IM,JM
      WRITE(1,107)H
      WRITE(1,108)AMAG
      WRITE(1,109)ALPHA
  105 FORMAT('REGION SIZE=',I2,'X',I2)
  106 FORMAT('MAGNET SIZE=',I2,'X',I2)
  107 FORMAT('MESH INTERVAL=',F5.2)
  108 FORMAT('MAGNETIZATION=',F5.2)
  109 FORMAT('CONVERGENCE FACTOR=',F5.2)
      WRITE(1,221)
      WRITE(1,221)
      WRITE(1,222)
  222 FORMAT(5X,'MESH USED')
      WRITE(1,221)
      EPMAX=1.0E-04
      EPSUM=1.0E-04
      HAM=H*AMAG
C   PRINT MESH PATTERN
      CALL MESH(IN,JN,IM,JM)
      INL=IN-1
      JNL=JN-1
C   SET OUTER BOUNDARIES
      DO 1 J=1,JN
      PHI(1,J)=0.0
    1 PHI(IN,J)=0.0
      DO 2 I=1,IN
      PHI(I,1)=0.0
    2 PHI(I,JN)=0.0
C   SET TRIAL INITIAL VALUES
      DO 5 I=2,INL
      DO 6 J=2,JNL
    6 PHI(I,J)=0.5
    5 CONTINUE
      ALPF=ALPHA/4.0
C   BEGIN ITERATION LOOP
      NCOUN=0
   10 NCOUN=NCOUN+1
      IF(NCOUN.GT.100) GOTO 11
      REMAX=0.0
      RESUM=0.0
      DO 3 J=1,JNL
      DO 4 I=1,INL
      IF((I.EQ.1).AND.(J.EQ.1)) GOTO 26
      IF((I.EQ.1).AND.(J.EQ.JM)) GOTO 23
      IF(I.EQ.1) GOTO 20
      IF(J.EQ.1) GOTO 21
      IF((I.EQ.IM).AND.(J.LT.JM)) GOTO 22
      IF((J.EQ.JM).AND.(I.LT.IM)) GOTO 23
      IF((J.EQ.JM).AND.(I.EQ.IM)) GOTO 24
```

PERMANENT MAGNETS

```
          R=PHI(I+1,J)+PHI(I-1,J)+PHI(I,J-1)+PHI(I,J+1)-4.0*PHI(I,J)
          GOTO 25
       20 R=2.0*PHI(2,J)+PHI(1,J-1)+PHI(1,J+1)-4.0*PHI(1,J)
          GOTO 25
       21 R=PHI(I+1,1)+PHI(I-1,1)-2.0*PHI(I,1)
          GOTO 25
       22 R=2.0*(PHI(IM+1,J)+PHI(IM-1,J))-4.0*PHI(IM,J)
          GOTO 25
       23 R=2.0*(PHI(I,JM+1)+PHI(I,JM-1)+HAM)-4.0*PHI(I,JM)
          GOTO 25
       24 R=PHI(IM+1,JM)+PHI(IM-1,JM)+PHI(IM,JM+1)+PHI(IM,JM-1)+
         1HAM-4.0*PHI(IM,JM)
          GOTO 25
       26 R=0.0
       25 CONTINUE
          PHI(I,J)=PHI(I,J)+ALPF*R
          AR=ABS(R)
          IF(AR.GT.REMAX) REMAX=AR
          RESUM=RESUM+R*R
        4 CONTINUE
        3 CONTINUE
C     TEST CONVERGENCE
          RESUM=SQRT(RESUM/FLOAT(INL*JNL))
          IF(REMAX.LT.EPMAX) GOTO 12
          IF(RESUM.GT.EPSUM) GOTO 10
       12 CONTINUE
       11 CONTINUE
          WRITE(1,120)NCOUN,REMAX,RESUM
      120 FORMAT(1H ,' NO OF ITERATIONS =',I3,'  REMAX= ',
         1E10.4,'     RESUM=',E10.4)
          WRITE(1,221)
          WRITE(1,221)
          WRITE(1,121)
      121 FORMAT(1H ,15X,'MAGNETOSTATIC POTENTIAL',)
          WRITE(1,221)
      221 FORMAT(1H )
C     CALCULATE MAGNETIC FIELD AND OUTPUT ARRAYS
          CALL OUT(IN,JN,PHI,9,3)
          CALL THEMA(IN,JN,H,PHI,TH,FE)
          WRITE(1,122)
      122 FORMAT(15X,'MAGNETIC FIELD DIRECTIONS',)
          WRITE(1,221)
          CALL OUT(IN,JN,TH,9,2)
          WRITE(1,123)
      123 FORMAT(15X,'MAGNETIC FIELD MAGNITUDES',)
          WRITE(1,221)
          CALL OUT(IN,JN,FE,9,2)
      111 FORMAT(1H ,SE10.2)
          STOP
          END
          SUBROUTINE OUT(IN,JN,F,NPRIN,NCON)
          DIMENSION F(50,50)
C     REORDERS ARRAY F AND PRINTS NPRIN COLUMNS AT A TIME
C     IF NCON=1 OUTPUTS CHARACTERS FOR MESH
          IP=0
        1 IP=IP+NPRIN
```

PERMANENT MAGNETS

```
      INP=IN
      IF( IP.LT.IN) INP=IP
      IL=IP-NPRIN+1
      JLN=1
      DO 2 JT=JLN,JN
      J=JN-JT+1
      IF(NCON.EQ.1) GOTO 3
      WRITE(1,111)(F(IK,J),IK=IL,INP)
      GOTO 2
    3 WRITE(1,113)(F(IK,J),IK=IL,INP)
    2 CONTINUE
      WRITE(1,112)
      WRITE(1,112)
      WRITE(1,112)
      IF(IP.LT.IN) GOTO 1
  111 FORMAT(1H ,12F7.2)
  112 FORMAT(1H )
  113 FORMAT(1H ,20A4)
      RETURN
      END
      SUBROUTINE MESH(IN,JN,IM,JM)
      DIMENSION Q(50,50)
      INTEGER T1,T2
C     SETS ARRAY OF CHARACTERS FOR MESH REGIONS
      DATA T1,T2/1H.,1H1/
      DO 1 I=1,IN
      DO 2 J=1,JN
      Q(I,J)=T1
    2 IF((I.LE.IM).AND.(J.LE.JM)) Q(I,J)=T2
    1 CONTINUE
      CALL OUT(IN,JN,Q,20,1)
      RETURN
      END
      SUBROUTINE THEMA(IN,JN,H,F,T,HM)
      DIMENSION F(50,50),HM(50,50),T(50,50)
C     INPUTS POTENTIAL AS F AND OUTPUTS MAGNETIC FIELD IN
C     ARRAYS T(DIRECTIONS) AND HM(MAGNITUDES)
      DO 1 I=2,IN
      DO 2 J=2,JN
      HX=-(F(I,J)-F(I-1,J))/H
      HY=-(F(I,J)-F(I,J-1))/H
      IF(HX.EQ.0.0) HX=0.01
      HXY=HY/HX
      TT=ATAN(ABS(HXY))*180.0/3.14159
C     ADJUST ARCTAN FOR DIFFERENT QUADRANTS
      IF((HX.GE.0.0).AND.(HY.GE.0.0)) P=TT
      IF((HX.LT.0.0).AND.(HY.GE.0.0)) P=180.0-TT
      IF((HX.LE.0.0).AND.(HY.LT.0.0)) P=180.0+TT
      IF((HX.GE.0.0).AND.(HY.LT.0.0)) P=360.0-TT
      T(I,J)=P
      HM(I,J)=SQRT(HX*HX+HY*HY)
    2 CONTINUE
    1 CONTINUE
      DO 3 J=2,JN
      HY=-(F(1,J)-F(1,J-1))/H
      T(1,J)=90.0
```

PERMANENT MAGNETS

```
   IF(HY.LT.0.0) T(1,J)=270.0
3 HM(1,J)=ABS(HY)
   DO 4 I=1,IN
   T(I,1)=270.0
4 HM(1,1)=HM(I,2)
   T(IN,1)=0.0
   RETURN
   END
```

PERMANENT MAGNETS RESULTS

REGION SIZE= 9X10
MAGNET SIZE= 4X 5
MESH INTERVAL= 0.10
MAGNETIZATION= 5.00
CONVERGENCE FACTOR= 1.50

   MESH USED

```
. . . . . . . .
. . . . . . . .
. . . . . . . .
. . . . . . . .
. . . . . . . .
1 1 1 1 . . . . .
1 1 1 1 . . . . .
1 1 1 1 . . . . .
1 1 1 1 . . . . .
1 1 1 1 . . . . .
```

NO OF ITERATIONS = 28  REMAX= 0.2361E-03      RESUM=0.9454E-04

## MAGNETOSTATIC POTENTIAL

```
0.00   0.00   0.00   0.00   0.00   0.00   0.00   0.00   0.00
0.14   0.14   0.13   0.11   0.08   0.06   0.04   0.02   0.00
0.30   0.29   0.26   0.22   0.17   0.12   0.08   0.04   0.00
0.47   0.45   0.41   0.34   0.26   0.18   0.11   0.05   0.00
0.66   0.65   0.59   0.48   0.33   0.22   0.13   0.06   0.00
0.90   0.88   0.82   0.65   0.38   0.23   0.14   0.06   0.00
0.63   0.61   0.55   0.43   0.30   0.20   0.12   0.06   0.00
0.40   0.38   0.34   0.27   0.21   0.14   0.09   0.04   0.00
0.19   0.19   0.17   0.13   0.10   0.07   0.05   0.02   0.00
0.00   0.00   0.00   0.00   0.00   0.00   0.00   0.00   0.00
```

*j* ↑

└──→ *i*

## MAGNETIC FIELD DIRECTIONS

```
 90.00   89.59   89.55   89.47   89.33   89.07   88.54   86.99    0.00
 90.00   88.36   84.74   80.29   74.81   68.41   59.39   41.49    0.00
 90.00   86.96   80.14   71.18   59.51   48.09   36.15   20.81    0.00
 90.00   85.96   76.82   63.07   42.18   28.34   18.73    9.87    0.00
 90.00   85.63   76.12   57.47   17.95    7.57    3.70    1.62    0.00
270.00  273.85  282.27  306.79  344.04  346.72  349.77  354.07    0.00
270.00  274.63  286.28  309.11  321.70  330.51  337.35  346.67    0.00
270.00  273.77  282.85  296.28  304.04  313.13  322.43  336.79    0.00
270.00  272.02  276.75  283.16  286.97  293.02  300.97  317.54    0.00
270.00  270.00  270.00  270.00  270.00  270.00  270.00  270.00    0.00
```

## MAGNETIC FIELD MAGNITUDES

```
1.44   1.40   1.27   1.08   0.85   0.61   0.39   0.19   0.01
1.52   1.48   1.35   1.14   0.89   0.64   0.43   0.27   0.19
1.70   1.65   1.52   1.28   0.98   0.74   0.55   0.42   0.37
1.97   1.93   1.82   1.54   1.16   0.90   0.71   0.58   0.52
2.32   2.33   2.38   2.05   1.53   1.15   0.87   0.70   0.62
2.68   2.68   2.76   2.81   2.82   1.50   0.98   0.74   0.64
2.31   2.27   2.15   1.96   1.57   1.19   0.86   0.65   0.56
2.05   1.99   1.82   1.56   1.23   0.94   0.68   0.50   0.41
1.92   1.86   1.67   1.38   1.08   0.79   0.53   0.32   0.22
1.92   1.86   1.67   1.38   1.08   0.79   0.53   0.32   0.22
```

Physics Programs
Edited by A. D. Boardman
© 1980 John Wiley & Sons Ltd.

CHAPTER 5

# Particle Capture in High Gradient Magnetic Separation

R. GERBER

## 1. INTRODUCTION

High gradient magnetic separation (HGMS) is a technique for the removal of weakly magnetic particles from suspensions that has many practical applications such as sewage and water treatments and the processing of industrial slurries. The method is based on the utilization of a magnetic traction force which extracts the particles from the fluid when the suspension passes through the separator system. The magnetic traction force, $\mathbf{F}_m$, is proportional to the difference, $\chi_p - \chi_f$, between the particle and fluid susceptibilities and, since curl $\mathbf{H} = 0$, to the product, $H$ grad $H$, of the magnitude of the magnetic field and its gradient at the position of the particle. The difference $\chi_p - \chi_f$ is usually very small for weakly magnetic particles, and also the magnetic field magnitude $H$ cannot be increased above a certain upper limit for technical reasons. Thus an efficient extraction, which results from a large value of $\mathbf{F}_m$, requires that the value of grad $H$ be high. Hence the name of the method.

An example of a HGMS system is shown schematically in Figure 1. It consists of a canister, filled with an ordered matrix of very thin magnetic wires, which is placed in a magnetic field, large enough to saturate the wires. The canister, using an appropriate plumbing, conducts either the suspension of mixed magnetic and non-magnetic particles or flush water through the system. The thin wires of the matrix dehomogenize the background magnetic field and give rise to a high value of its gradient in their immediate surroundings. When the magnetic field is on, the suspension of particles to be separated flows through the matrix, the magnetic particles are captured on to the wires, and the purified suspension leaves the system. At intervals, when the retention capacity of the matrix is reached, the feed is halted, the magnetic field is switched off, and the magnetic particles are flushed out of the separator. Then the cycle is repeated.

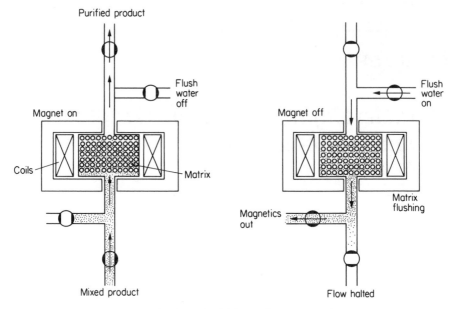

Figure 1. Schematic view of a high gradient magnetic separator

The packing fraction, $F$ (which is defined as the volume occupied by the wires to that of the canister), of the matrix is typically between 0.05 and 0.1. Thus most of the space in the canister is actually free. This has two important consequences. First, the fluid impedance of the system is low. Hence a large amount of material can be passed through and treated by the separator in a short period of time. Second, the capture process can be considered in a single-wire approximation.[1] In this approximation, since the wires in the canister are relatively far apart, it is assumed that the particles interact with only one individual wire at a time.

An important quantity characterizing any separation system is the filter performance. It is given as the ratio $N_{out}/N_{in}$, where $N_{out}$ and $N_{in}$ are the number of magnetic particles leaving and entering the separator, respectively. The smaller the ratio the better the performance.

In order to obtain some idea of filter performance let us consider an ordered matrix comprised of a large number of stacked-up sheets of gauze which are woven in a regular rectangular pattern from a thin ferromagnetic wire of radius $a$. If the magnetic field is normal to the surface of each sheet and if certain limiting conditions are fulfilled, it can be shown[2] that the filter performance is given by the formula

$$N_{out} = \exp\left(\frac{-2R_{ca}FL}{\pi a}\right)N_{in}, \qquad (1)$$

where $L$ is the total length of the matrix stack and $R_{ca}$ is the capture radius (in units of $a$) associated with an individual wire.

It is clear that particle trajectories and capture radii are of fundamental importance for HGMS. A useful starting point for an investigation of this technique is a computational study dealing with particle capture in the single-wire approximation, and this is presented in this chapter.

## 2. SINGLE-WIRE APPROXIMATION

### 2.1 Configuration

The background magnetic field in the separator becomes inhomogeneous due to the appearance of magnetic charges on the ferromagnetic wires of the matrix. This effect is a maximum if the ordered matrix is arranged so that the direction of the field is always perpendicular to the axes of the wires. Therefore, in the single-wire approximation, we consider a configuration as shown in Figure 2. A ferromagnetic wire of radius $a$, saturation magnetization $M_s$, is placed axially along the $z$-axis of an orthogonal coordinate system. A magnetic field, $H_0$, sufficient to saturate the wire, is applied in the $x$-direction and a fluid of viscosity $\eta$ flows with a background velocity $V_0$ along a direction which lies in the $xy$ plane making an angle $\alpha$ with the $x$-axis.

A completely arbitrary direction of fluid flow is perfectly possible in

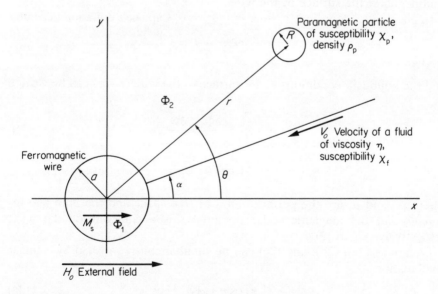

Figure 2. Configuration of the particle capture problem in the single-wire approximation

HGMS. However, such direction can be resolved in one lying in the $xy$ plane and in one parallel with the $z$-axis. The latter case, the so-called axial configuration, can be solved analytically[3] and is therefore outside the scope of this chapter.

Paramagnetic particles of spherical shape with radius $R (R \ll a)$, volume $V_p$, susceptibility $\chi_p$ and density $\rho_p$, are carried by the fluid. The flow of the fluid around the wire is considered to be streamlined and frictionless, but the particles are assumed to experience a viscous drag force $\mathbf{F}_v$ given by the Stokes equation $\mathbf{F}_v = -6\pi\eta R(\mathbf{v} - \mathbf{V})$, where $(\mathbf{v} - \mathbf{V})$ is the relative velocity of the particle with respect to the fluid.

Physical quantities used throughout this chapter are in SI units (Sommerfeld system). The metric part of their dimensions may be sometimes expressed in terms of the wire radius $a$. Then, to indicate this renormalization, the symbols are provided with the subscript $a$ e.g. $R_a = R/a$, $V_{pa} = V_p/a^3$, etc.).

## 2.2 Magnetic field

The magnetic field $\mathbf{H}$ of a wire magnetized by a uniform background field $\mathbf{H}_0$ can be expressed as $\mathbf{H} = -\operatorname{grad}\Phi$, where $\Phi$ is the magnetic scalar potential. Since $\operatorname{div}\mathbf{B} = 0$, we can write for a linear medium $\operatorname{div}\mathbf{B} = \operatorname{div}(\mu\mathbf{H}) = \mu \operatorname{div}\mathbf{H} = \mu \operatorname{div}\operatorname{grad}\Phi = 0$. Thus the potential $\Phi$ must satisfy Laplace's equation $\nabla^2\Phi = 0$ and also the boundary conditions requiring the tangential components of $\mathbf{H}$ and the normal components of $\mathbf{B}$ to be continuous at the surface of the wire.

In plane polar coordinates, since $\partial^2\Phi/\partial z^2 = 0$, Laplace's equation becomes

$$r\frac{\partial}{\partial r}\left(r\frac{\partial\Phi}{\partial r}\right) + \frac{\partial^2\Phi}{\partial\theta^2} = 0 \tag{2a}$$

and the boundary conditions at the surface of the wire ($r = a$) can be written as

$$\frac{\partial\Phi_1}{\partial\theta} = \frac{\partial\Phi_2}{\partial\theta}, \tag{2b}$$

$$\mu_0\left(-\frac{\partial\Phi_1}{\partial r} + M_s\cos\theta\right) = -\mu_f\frac{\partial\Phi_2}{\partial r}, \tag{2c}$$

where $\mu_0$ and $\mu_f$ are the permeabilities of free space and the fluid and $\Phi_1$ and $\Phi_2$ are the magnetic scalar potentials inside and outside the wire, respectively (Figure 2).

Equations (2a), (2b), and (2c) can be simultaneously satisfied by a linear combination

$$\Phi_1 = -C_1 r\cos\theta + A_1 r^{-1}\cos\theta, \tag{3a}$$

$$\Phi_2 = -C_2 r\cos\theta + A_2 r^{-1}\cos\theta \tag{3b}$$

of cylindrical harmonics.[4]

The field in the origin (on the wire axis) is finite, hence $A_1 = 0$. At large distances, the field tends to $H_0$, consequently $C_2 = H_0$. The remaining two constants, $C_1$ and $A_2$, are obtained by substitution of equations (3a, b) into (2b, c). We obtain

$$C_1 = H_0 - A_2 a^{-2} \tag{4}$$

and

$$A_2 = \frac{\mu_0 - \mu_f}{\mu_0 + \mu_f} H_0 a^2 + \frac{\mu_0}{\mu_0 + \mu_f} M_s a^2. \tag{5}$$

Since for most practical cases $|\mu_0 - \mu_f| \lesssim 10^{-3}$, the first term in equation (5) can be neglected in comparison with the second. Thus we obtain from (4) and (5)

$$A_2 = \tfrac{1}{2} M_s a^2, \tag{6}$$

$$C_1 = H_0 - \tfrac{1}{2} M_s. \tag{7}$$

Substituting $A_1 = 0$, $C_2 = H_0$, equations (6) and (7) into (3a, b) we obtain the scalar potentials, the negative gradient of which gives the respective magnetic fields, $\mathbf{H}_1$ and $\mathbf{H}_2$, inside and outside the wire. Their components are

$$H_{1r} = (H_0 - \tfrac{1}{2} M_s)\cos \theta, \tag{8a}$$

$$H_{1\theta} = -(H_0 - \tfrac{1}{2} M_s)\sin \theta, \tag{8b}$$

and

$$H_{2r} = (\tfrac{1}{2} M_s a^2 r^{-2} + H_0)\cos \theta, \tag{9a}$$

$$H_{2\theta} = (\tfrac{1}{2} M_s a^2 r^{-2} - H_0)\sin \theta. \tag{9b}$$

## 2.3 Fluid velocity distribution

The fluid is assumed to be viscous when interacting with the particles, but when interacting with the wires of the filter it is considered to be ideal. Consequently, if irrotational flow is assumed, the velocity of the fluid motion is given as $\mathbf{V} = -\text{grad } \psi$, where $\psi$ is the velocity potential. Again, the potential $\psi$ must satisfy Laplace's equation $\nabla^2 \psi = 0$ and also the boundary condition that the normal component of the velocity must vanish at the surface of the wire.

Therefore, in plane polar coordinates, $\psi$ has to satisfy simultaneously equation (2a), when substituted instead of $\Phi$, and the condition $\partial \psi / \partial r = 0$ for $r = a$ and any $\theta$. Bearing in mind that at large distances the flow takes place at an angle $\alpha$ with a uniform velocity $V_0$, we can see immediately that the function

$$\psi = -V_0 r \cos(\theta - \alpha) + A r^{-1} \cos(\theta - \alpha), \tag{10}$$

which is analogous to (3a, b), will satisfy both equation (2a) and the boundary condition. Substitution of (10) into $(\partial\psi/\partial r)_{r=a}=0$ gives $A = -V_0a^2$. This value can now be used in (10), when the negative gradient is taken of this expression. Thus we obtain the components of the fluid velocity:

$$V_r = V_0(1 - a^2r^{-2})\cos(\theta - \alpha), \tag{11a}$$

$$V_\theta = -V_0(1 + a^2r^{-2})\sin(\theta - \alpha). \tag{11b}$$

## 2.4 Equations of particle motion

The equation of motion can be written in a vector form as

$$m\mathbf{a} = \mathbf{F}_v + \mathbf{F}_m, \tag{12}$$

where $m = \frac{4}{3}\pi R^3\rho_p$ is the mass of the particle, $\mathbf{a}$ is the acceleration, $\mathbf{F}_v$ and $\mathbf{F}_m$ are the viscous drag and magnetic traction force, respectively. In order to be able to solve equation (12) we need to establish the components of all its terms in plane polar coordinates $r$ and $\theta$.

Consider two mutually perpendicular unit vectors $\hat{\mathbf{r}}$ and $\hat{\boldsymbol{\theta}}$ which have directions of increasing $\mathbf{r}$ and increasing $\theta$. Their directions change with time and since the derivative of a unit vector is perpendicular to the vector, we have

$$\frac{d\hat{\mathbf{r}}}{dt} = \frac{d\theta}{dt}\hat{\boldsymbol{\theta}}, \tag{13a}$$

$$\frac{d\hat{\boldsymbol{\theta}}}{dt} = -\frac{d\theta}{dt}\hat{\mathbf{r}}. \tag{13b}$$

Now, the position of the particle in plane polar coordinates is given by $\mathbf{r} = r\hat{\mathbf{r}}$. Differentiating this equation and using (13a, b), we obtain the particle velocity $\mathbf{v}$, the components of which are $v_r = dr/dt$ and $v_\theta = r(d\theta/dt)$. By second differentiation, we obtain the particle acceleration $\mathbf{a}$ and its components

$$a_r = \frac{d^2r}{dt^2} - r\left(\frac{d\theta}{dt}\right)^2, \tag{14a}$$

$$a_\theta = r\frac{d^2\theta}{dt^2} + 2\frac{dr}{dt}\frac{d\theta}{dt}. \tag{14b}$$

The components of the viscous drag force $\mathbf{F}_v$ are

$$F_{vr} = -6\pi\eta R\left(\frac{dr}{dt} - V_r\right) \tag{15a}$$

and

$$F_{v\theta} = -6\pi\eta R\left(r\frac{d\theta}{dt} - V_\theta\right), \tag{15b}$$

where $V_r$ and $V_\theta$ are given by equations (11a, b).

To establish the components of the magnetic traction force $\mathbf{F}_m$ we have to consider first the magnetic energy of a system comprising a fluid and a particle.

The magnetic energy density is expressed in general as $\frac{1}{2}HB$. Imagine a volume $V_p$ demarcated inside the fluid. The magnetic energy of the fluid enclosed in this volume is $\frac{1}{2}HB_f V_p = \frac{1}{2}\mu_f V_p H^2$. Let us now remove the fluid from the volume $V_p$ and replace it by the particle. The energy associated with the particle itself is $\frac{1}{2}\mu_p V_p H^2$. The energy increment $U$ of the system (particle + fluid) is given as the difference between these two energies, i.e. $U = \frac{1}{2}(\mu_p - \mu_f)V_p H^2$. Taking $\nabla U$ (positive gradient, since the magnetic energy is of the same nature as the kinetic energy) we get

$$\mathbf{F}_m = \frac{1}{2}\mu_0 \chi V_p \nabla(H^2), \tag{16}$$

where $\chi = \chi_p - \chi_f$ is the difference between susceptibilities of the particle and the fluid.

Considering a spherical particle, $V_p = \frac{4}{3}\pi R^3$, and substituting $H^2 = H_{2r}^2 + H_{2\theta}^2$ from (9a, b) into (16), we obtain, after some manipulation, the components of the magnetic traction force

$$F_{mr} = -\frac{4\pi\mu_0\chi M_s a^2 R^3}{3r^3}\left(\frac{M_s a^2}{2r^2} + H_0\cos 2\theta\right), \tag{17a}$$

$$F_{m\theta} = -\frac{4\pi\mu_0\chi M_s a^2 R^3}{3r^3}H_0\sin 2\theta. \tag{17b}$$

Combining equations (14a, b), (15a, b), and (17a, b) with (12) we obtain the complete equations of motion: in radial direction

$$\frac{2\rho_p R^2}{9\eta}\left(\frac{d^2 r_a}{dt^2} - r_a\left(\frac{d\theta}{dt}\right)^2\right) + \frac{dr_a}{dt}$$

$$= V_{0a}\left(1 - \frac{1}{r_a^2}\right)\cos(\theta - \alpha) - V_{ma}\left(\frac{M_s}{2H_0 r_a^2} + \cos 2\theta\right)\frac{1}{r_a^3} \tag{18a}$$

and in azimuthal direction

$$\frac{2\rho_p R^2}{9\eta}\left(r_a\frac{d^2\theta}{dt^2} + 2\frac{dr_a}{dt}\frac{d\theta}{dt}\right) + r_a\frac{d\theta}{dt}$$

$$= -V_{0a}\left(1 + \frac{1}{r_a^2}\right)\sin(\theta - \alpha) - V_{ma}\frac{\sin 2\theta}{r_a^3}, \tag{18b}$$

**where** $r_a = r/a$, $V_{0a} = V_0/a$, $V_{ma} = V_m/a = \frac{2}{9}(\chi\mu_0 M_s H_0 R_a^2/\eta)$ and $R_a = R/a$. The quantity $V_m$, the dimensions of which are m s$^{-1}$, is called the magnetic velocity.

The first (inertial) terms in equations (18a, b) can usually be neglected in comparison with others in case of HGMS of small particles in liquids. Then $R^2$ is very small and $\eta$ moderately high and (18a, b) reduce to a pair of first-order differential equations:

$$\frac{dr_a}{dt} = V_{0a}\left(1 - \frac{1}{r_a^2}\right)\cos(\theta - \alpha) - V_{ma}\left(\frac{K}{r_a^2} + \cos 2\theta\right)\frac{1}{r_a^3}, \tag{19a}$$

$$\frac{d\theta}{dt} = -V_{0a}\left(1 + \frac{1}{r_a^2}\right)\frac{\sin(\theta - \alpha)}{r_a} - V_{ma}\frac{\sin 2\theta}{r_a^4}, \tag{19b}$$

where $K = M_s/2H_0$. Equations (19a, b) will be analysed and numerically solved in the main text of this chapter.

If the particle bearing fluid is a gas, $\eta$ is rather small and the inertial term cannot be neglected. Complete equations of motion are therefore required for a numerical solution of this problem.[5] An appropriate extension of our computer programs to perform this task also is included in the suggested exercises.


## 3. NUMERICAL SOLUTION OF ORDINARY DIFFERENTIAL EQUATIONS

### 3.1 Predictor–corrector method

Consider a first-order initial-value problem

$$y' = f(x, y), \tag{20a}$$

$$y(x_0) = y_0, \tag{20b}$$

where $x_0$, $y_0$, and the function $f(x, y)$ are given, $y'$ denotes d$y$/d$x$. Under quite general conditions this problem has a unique solution $y(x)$ in the interval[6] $\langle x_0, \bar{x}\rangle$, $\bar{x} \neq x_0$.

We would like to find a numerical approximation to the solution $y(x)$. Let us divide the interval $\langle x_0, \bar{x}\rangle$ into $m$ subintervals, each of length $h$. Thus, there are $m + 1$ points

$$x_n = x_0 + nh, \qquad n = 0, 1, 2, \ldots, m,$$

where we are seeking the values $y_1, y_2, \ldots, y_m$ of the solution $y(x)$.

The numerical method of solving differential equations such as (20a, b) involves the procedure which yields $y_{n+1}$, given the sequence $y_n, y_{n-1}, y_{n-2} \ldots$ up to $y_{n-s}$. The value of $s$ may be 0, 1, 2, etc. When $y_{n+1}$ is found, the procedure is repeated until the whole interval $\langle x_0, \bar{x}\rangle$ is covered.

The Euler method, which has $s = 0$, is the simplest numerical method for solving initial-value problems of the type (20a, b). To derive this method, assume that the value $y_n$ is known and integrate (20a) from $x_n$ to $x_{n+1}$. We get

$$y_{n+1} - y_n = \int_{x_n}^{x_{n+1}} f(x, y(x)) \, dx. \tag{21}$$

The integral in (21) can be approximated by the rectangle rule as $hf(x_n, y(x_n))$. Thus we obtain the formula (procedure)

$$y_{n+1} - y_n = hf(x_n, y_n). \tag{22}$$

Starting with $n = 0$, i.e. substituting first the known value $y_0$, we find by repetition all required values.

Evidently, the accuracy of determination of $y_{n+1}$ depends on $h$. It can be shown that the single step (quadrature) error is approximately proportional to $h^2$ for the Euler method. We can, therefore, increase the accuracy by decreasing the size of $h$. This will, however, increase the number of steps and hence the computing time. Moreover, with each step an accumulated error will enter into the process. This error will partly offset the improvement in accuracy which was gained by the reduction of $h$. Clearly, one would like to increase the accuracy without the need to reduce $h$. This can be achieved by using numerical quadrature formulae, based on Lagrangian interpolation, for the approximation of the integral in equation (21). These quadrature formulae take into account the behaviour of the function $f(x, y(x))$ in a wider surroundings of the interval of integration. Thus the integral in (21) can be expressed to a high degree of accuracy for a relatively large $h$.

We have two types of formulae available. The open or predictor formula

$$y_{n+1} - y_{n+1-q} = h \sum_{i=1}^{M} \beta_i f(x_{n+1-i}, y_{n+1-i}), \tag{23a}$$

and the closed or corrector formula

$$y_{n+1} - y_{n+1-q} = h \sum_{i=0}^{M'} \beta_i' f(x_{n+1-i}, y_{n+1-i}), \tag{23b}$$

where $q$ is a fixed integer, and $\beta_i$ and $\beta_i'$ are the numerical coefficients. A full analysis and values of $q$, $\beta_i$, and $\beta_i'$ of various methods can be found in Young and Gregory.[6]

Note that equation (23a) contains $y_{n+1}$ explicitly, whereas (23b) contains $y_{n+1}$ implicitly, hence the names 'open' and 'closed' formula, respectively.

Formulae (23a, b) together form the predictor–corrector method. It works as follows: let us assume that we are at the $n$th point and that we already know, by some means, the values $y_n, y_{n-1}, \ldots, y_{n-M+1}$. Substituting these known values in (23a) we obtain the predicted value $y^{(P)}$. Substituting $y^{(P)}_{n+1}$ together with $y_n, y_{n-1}, \ldots$ etc. in (23b) we obtain the corrected value $y^{(C)}_{n+1}$. If $|(y^{(P)}_{n+1} - y^{(C)}_{n+1})/y^{(C)}_{n+1}| < E$, where $E$ is the stipulated error, we put $y_{n+1} = y^{(C)}_{n+1}$ and repeat the cycle for the next point, i.e. for $n+1$. If the relative error between the predicted and corrected value is larger than $E$ we have to reduce $h$ and calculate the next $M-1$ values by some other means (see section 3.2) before the predictor–corrector method can again be employed.

There is a large family of predictor–corrector methods available. The Adams–Moulton method, which is employed in our program, is particularly attractive, since it is of a good stability and high precision (quadrature error $\approx 0(h^5)$). The numerical quantities characterizing this method are $q = 1$, $M = 4$, $\beta_i = 55/24$, $-59/24$, $37/24$, $-9/24$, $M' = 3$, $\beta'_i = 9/24$, $19/24$, $-5/24$, $1/24$.

## 3.2 Runge–Kutta method

To get a predictor–corrector method into action we need to know the values of $y$ at several starting points $x_0 + h$, $x_0 + 2h$, $\ldots$, $x_0 + sh$, in addition to the given value of $y(x_0)$. Hence the desire for an accurate self-starting method, i.e. a high order method which would require a knowledge of $y$ at only one initial point. The Runge–Kutta method is of this type and its derivation will be briefly described here. Consider again equation (21). We can write

$$y^{(P)}_{n+1} = y_n + hf(x_n, y_n), \tag{24a}$$

$$y_{n+1} = y_n + \frac{h}{2}[f(x_n, y_n) + f(x_{n+1}, y^{(P)}_{n+1})], \tag{24b}$$

where the predicted value $y^{(P)}_{n+1}$ and the corrected value $y_{n+1}$ were obtained by applying the rectangle and the trapezoidal rule, respectively. Formulae (24a, b), which constitute the Heun predictor–corrector method, can be rewritten in the following form:

$$y_{n+1} = y_n + \alpha_1 \Delta_1 + \alpha_2 \Delta_2, \tag{25}$$

where

$$\Delta_1 = hf(x_n, y_n), \tag{26a}$$

$$\Delta_2 = hf(x_n + \rho h, y_n + \rho \Delta_1), \tag{26b}$$

and where the numerical coefficients have the values of $\rho = 1$, $\alpha_1 = \frac{1}{2}$, $\alpha_2 = \frac{1}{2}$.

Thus instead of two formulae, (24a, b), we have one formula, (25), with two 'differential' increments $\Delta_1$ and $\Delta_2$. The first increment $\Delta_1$ serves not only in (25) but also as a 'predictor' for the second increment $\Delta_2$ in (26b).

Generalizing formulae (25) and (26a, b) to involve four increments and determining the appropriate numerical coefficients by comparison of $y_{n+1} = \sum_{j=1}^{4} \alpha_j \Delta_j$ with $y_{n+1}$ given as a Taylor's series expansion of $y(x)$ at $x_n$, we obtain the (fourth-order) Runge–Kutta method as

$$y_{n+1} = y_n + \tfrac{1}{6}(\Delta_1 + 2\Delta_2 + 2\Delta_3 + \Delta_4), \tag{27}$$

where

$$\Delta_1 = hf(x_n, y_n), \tag{28a}$$

$$\Delta_2 = hf(x_n + \tfrac{1}{2}h, y_n + \tfrac{1}{2}\Delta_1), \tag{28b}$$

$$\Delta_3 = hf(x_n + \tfrac{1}{2}h, y_n + \tfrac{1}{2}\Delta_2), \tag{28c}$$

$$\Delta_4 = hf(x_n + h, y_n + \Delta_3). \tag{28d}$$

The Runge–Kutta method, the single step error of which is $O(h^5)$, is used in our programs either in its own right or in connection with the Adams–Moulton method as a starter. This choice satisfies the rule that the starter should have approximately the same accuracy as the main execution method.

### 3.3 Systems of equations and higher order equations

Many physical problems involve coupled differential equations such as (18a, b) and (19a, b). Consequently, it is necessary to develop a method of solving them. Consider a first-order system

$$y_1'(x) = f_1(x, y_1, y_2, \ldots, y_n),$$
$$y_2'(x) = f_2(x, y_1, y_2, \ldots, y_n), \tag{29a}$$
$$\begin{array}{cc} \cdot & \cdot \\ \cdot & \cdot \\ \cdot & \cdot \end{array}$$
$$y_n'(x) = f_n(x, y_1, y_2, \ldots, y_n),$$

with initial conditions

$$y_1(x_0), y_2(x_0), \ldots, y_n(x_0), \tag{29b}$$

where $x_0$, initial conditions (29b), and the functions

$$f_j(x, y_1, y_2, \ldots, y_n), j = 1, 2, \ldots, n,$$

are given.

Let us introduce the vector notation:[6]

$$\mathbf{y}(x) = \begin{pmatrix} y_1(x) \\ y_2(x) \\ \cdot \\ \cdot \\ \cdot \\ y_n(x) \end{pmatrix}, \qquad \mathbf{y}'(x) = \begin{pmatrix} y_1'(x) \\ y_2'(x) \\ \cdot \\ \cdot \\ \cdot \\ y_n'(x) \end{pmatrix},$$

$$\mathbf{f}(x, \mathbf{y}) = \begin{pmatrix} f_1(x, y_1, y_2, \dots, y_n) \\ f_2(x, y_1, y_2, \dots, y_n) \\ \cdot \\ \cdot \\ \cdot \\ f_n(x, y_1, y_2, \dots, y_n) \end{pmatrix} = \begin{pmatrix} f_1(x, \mathbf{y}) \\ f_2(x, \mathbf{y}) \\ \cdot \\ \cdot \\ \cdot \\ f_n(x, \mathbf{y}) \end{pmatrix},$$

etc. The system (29a) and the initial conditions (29b) can then be rewritten as follows:

$$\mathbf{y}' = \mathbf{f}(x, \mathbf{y}), \tag{30a}$$

$$\mathbf{y}(x_0) = \mathbf{y}_0. \tag{30b}$$

Equations (30a, b) are of the same form as (20a, b). Consequently, the Adams–Moulton method for solution of a system of first-order differential equations can be obtained from (23a, b) if scalar quantities are replaced by vectors and the numerical coefficients $\beta_i$, $\beta_i'$ are left unchanged. Thus we get

$$\mathbf{y}_{n+1}^{(P)} = \mathbf{y}_n + \frac{h}{24}\, (55\mathbf{f}(x_n, \mathbf{y}_n) - 59\mathbf{f}(x_{n-1}, \mathbf{y}_{n-1})$$

$$+ 37\mathbf{f}(x_{n-2}, \mathbf{y}_{n-2}) - 9\mathbf{f}(x_{n-3}, \mathbf{y}_{n-3})), \tag{31a}$$

$$\mathbf{y}_{n+1} = \mathbf{y}_n + \frac{h}{24}\, (9\mathbf{f}(x_{n+1}, \mathbf{y}_{n+1}^{(P)}) + 19\mathbf{f}(x_n, \mathbf{y}_n)$$

$$- 5\mathbf{f}(x_{n-1}, \mathbf{y}_{n-1}) + \mathbf{f}(x_{n-2}, \mathbf{y}_{n-2})). \tag{31b}$$

Similarly, the Runge–Kutta method is obtained from (27) and (28a, b, c, d) as

$$\mathbf{y}_{n+1} = \mathbf{y}_n + \tfrac{1}{6}(\boldsymbol{\Delta}_n^{(1)} + 2\boldsymbol{\Delta}_n^{(2)} + 2\boldsymbol{\Delta}_n^{(3)} + \boldsymbol{\Delta}_n^{(4)}), \tag{32}$$

where

$$\boldsymbol{\Delta}_n^{(1)} = h\mathbf{f}(x_n, \mathbf{y}_n), \tag{33a}$$

$$\boldsymbol{\Delta}_n^{(2)} = h\mathbf{f}(x_n + \tfrac{1}{2}h, \mathbf{y}_n + \tfrac{1}{2}\boldsymbol{\Delta}_n^{(1)}), \tag{33b}$$

$$\boldsymbol{\Delta}_n^{(3)} = h\mathbf{f}(x_n + \tfrac{1}{2}h, \mathbf{y}_n + \tfrac{1}{2}\boldsymbol{\Delta}_n^{(2)}), \tag{33c}$$

$$\boldsymbol{\Delta}_n^{(4)} = h\mathbf{f}(x_n + h, \mathbf{y}_n + \boldsymbol{\Delta}_n^{(3)}). \tag{33d}$$

It is easy to show that the problem of solving a single differential equation of higher order can be reduced to the problem of solving a system of first-order differential equations.

Consider

$$y^{(n)} = f(x, y, y', \ldots, y^{(n-1)}), \tag{34a}$$

where

$$y(x_0) = y_0, \; y'(x_0) = y'_0, \ldots, y^{(n-1)}(x_0) = y_0^{(n-1)} \tag{34b}$$

are the given initial conditions at the point $x_0$.

We can define a set of new functions as

$$y_1(x) = y(x), \; y_2(x) = y'(x), \ldots, y_n(x) = y^{(n-1)}(x).$$

It is obvious that the functions $y_1(x), y_2(x), \ldots, y_n(x)$ satisfy the system

$$
\begin{aligned}
y'_1 &= y_2 \\
y'_2 &= y_3 \\
&\phantom{=}\;\vdots \\
y'_{n-1} &= y_n \\
y'_n &= f(x, y_1, y_2, \ldots, y_n),
\end{aligned}
\tag{35a}
$$

and the initial conditions

$$y_1(x_0) = y_0, \qquad y_2(x_0) = y'_0, \ldots, \qquad y_n(x_0) = y_0^{(n-1)}. \tag{35b}$$

Thus by solving (35a, b) we find the function $y_1(x) = y(x)$ and its $n$ derivative which are the solution to (34a, b).

Consequently, formulae (31a, b) or (32) and (33a, b, c, d) can also be used for finding a numerical solution of an equation of higher order if the indicated transformation to a system of first-order equations is used.

## 4. COMPUTER PROGRAMS

### 4.1 Main program

The objective of the main program is to calculate particle trajectories and capture radii $R_{ca}$ with a prescribed accuracy for a wide variety of input parameters and initial conditions. The program consists of the MASTER section and the subroutines FUNCTN, RK, AM and XSECT.

The MASTER section handles the input–output operations and calls the relevant subroutines to perform the computation.

The input data are:

J—the control integer which selects the iterating subroutine;

K—the control integer which selects whether the output is directed to a lineprinter or/and to a file;

IO—the number of trajectories or capture radii $R_{ca}$ to be calculated;

ALPHA—the angle $\alpha$ which the fluid flow makes with the external field $H_0$. These directions also enable us to define two convenient frames of reference, the primary has an $x$-axis parallel to $H_0$ and the secondary has an $x$-axis anti-parallel to the flow (Figure 3).

And

(a) if either the RK or AM subroutine has been chosen:

XAO and YAO—the starting point in the secondary frame of reference. These values are transformed in primary coordinates and

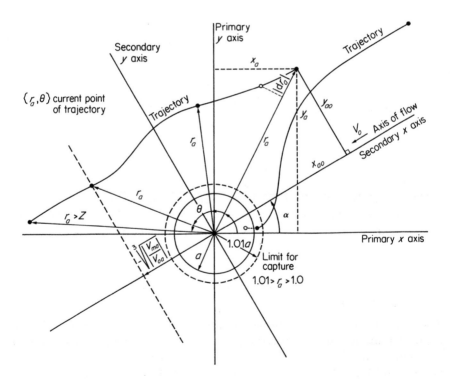

Figure 3. Definition of the primary and secondary frame of coordinates and the limiting conditions of capture and escape

later (by subroutines) in plane polar coordinates, where the iteration takes place;

VMA—the magnetic velocity $V_{ma}$;

VOA—the fluid velocity $V_{0a}$;

AK—the constant $K = M_s/2H_0$;

DTO—the time step length (a good empirical choice is DTO = $1/(10\ V_{0a})$), DTO is converted to the variable DT which can be adjusted during the iteration in subroutines;

E—the stipulated error (e.g. $E = 0.001$) of the Admas–Moulton predictor–corrector method.

Or

(b)  if the XSECT subroutine has been chosen:
VMA, VOA, AK, DTO, ALPHA, XAO;

YLA and YHA—the lower and upper estimate of the $y$-coordinate (in the secondary frame) of a point lying on the critical curve.

The output data are:

(a)  if either the RK or AM subroutine has been chosen:
AK, VMA, VOA, VMA/VOA, ALPHA, XAO, YAO;

X and Y—the $x$–$y$ coordinates of the trajectory points in the primary frame of reference;

V—the velocity of the particle at each trajectory point;

DAR and DIT—the radial and angular components, $dr_a/dt$ and $d\theta/dt$, of the velocity at each point;

DEET—the time step length $dt$ between the points;

(b)  if the XSECT subroutine has been chosen:
ITERATIONS and then AK, VMA, VOA, VMA/VOA, ALPHA;

XC—the $x$-coordinate (in the secondary frame) corresponding to the capture radius $R_{ca}$ on the critical curve;

YC—the capture radius $R_{ca}$.

The flow chart of the MASTER section is shown in Figure 4.

The subroutine FUNCTN contains the equations of motion in the form of (19a, b).

The subroutines RK and AM each transfer the input parameters and the starting point from the MASTER section, compute one trajectory, return it to the MASTER section (to be printed out), and receive a new set of initial conditions to calculate the next trajectory.

The RK and AM subroutines use the following limiting conditions:

(i) the particle is captured $\leftrightarrow 1.0 < r_a < 1.01$;

(ii) the particle is beyond the range of interest $\leftrightarrow r_a > Z = \max(14.2, r_{a0} + 0.2)$;

(iii) the impenetrability of the wire is violated $\leftrightarrow r_a < 1.0$;

(iv) the overall step limit is violated $\leftrightarrow |dr_a| > 0.1$.

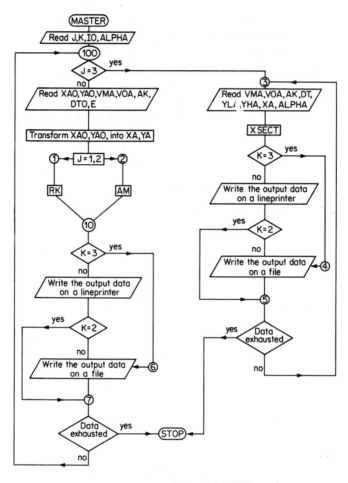

Figure 4. Flow chart of the MASTER section

If the condition (i) or (ii) is satisfied the control is returned to the MASTER section; if (iii) or (iv) takes place the step length DT is cut down.

The RK subroutine is based on the Runge–Kutta method as expressed by (32) and (33a, b, c, d). The AM subroutine uses the Runger–Kutta method as a starter and the Adams–Moulton method, given by (31a, b) as the main procedure. The role of the starter is not limited only to the initial point. After every cut in the step length DT the execution process is reversed to the Runge–Kutta method to provide four equally spaced points which are necessary for the action of the main predictor–corrector procedure. The process of execution in both subroutines, RK and AM, is apparent from the flow charts, which are shown in Figures 5 and 6, respectively.

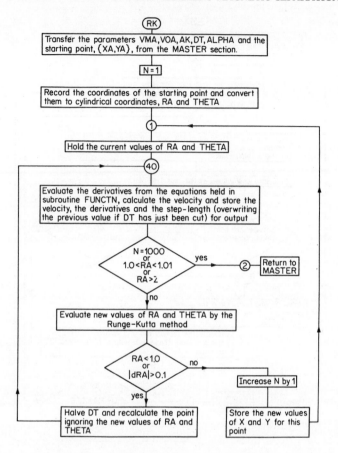

Figure 5. Flow chart of the RK subroutine

The XSECT subroutine calculates the capture radius $R_{ca}$. Again, the input data are transferred from the MASTER section, value of $R_{ca}$ is computed and returned with iterations to be printed out. A new set of input data is transferred and the cycle repeated.

Before describing the flow of control in XSECT we have to consider a few points of a general nature.

It is apparent from equations (11a, b), (15a, b) and (17a, b) that the influence which the wire has upon the motion of the particle falls off with the distance $r_a$. Thus, beyond a certain region, the dimensions of which increases with $|V_{ma}/V_{0a}|$, the effect of the wire can be neglected and the trajectories are essentially parallel to the direction of the fluid flow (i.e. to

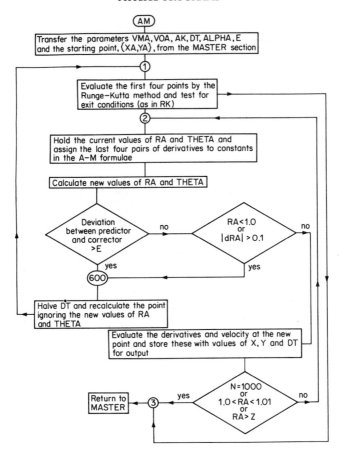

Figure 6. Flow chart of the AM subroutine

the secondary $x$-axis). The trajectories can be divided into two kinds, differing whether the particle is captured or not. The borderline case of a trajectory for which the particle is just captured is called the critical trajectory. The distance between the parallel part of the critical trajectory and the $x$-axis defines the capture radius (see Figure 7).

There are two critical trajectories, one above and one below the axis of flow. Thus we have two capture radii, $R_{cal}$ and $R_{ca2}$. The average capture radius $\frac{1}{2}(R_{cal} + R_{ca2})$ is exactly the quantity $R_{ca}$ which appears in equation (1). The proof that this single-wire analysis applies to the multiwire regular matrix can be found in Birss, Gerber, and Parker.[2]

The XSECT subroutine always computes in principle the capture radius, $R_{cal}$, above the axis of flow. To obtain $R_{ca2}$, the computation has to be

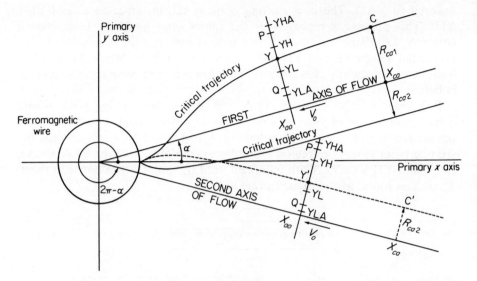

Figure 7.  Arrangement of the axes of flow and various parameters used in the XSECT subroutine

performed in the mirror image of the original flow with respect to the primary $x$-axis. This is achieved by the same input data as for $R_{cal}$, only instead of $\alpha$ an angle $\beta = 2\pi - \alpha$ is used.

The XSECT subroutine uses the same limiting conditions (i), (iii), and (iv) as the previous subroutines. However, instead of (ii) a more rigorous condition, which would be linked to the ratio $|V_{ma}/V_{0a}|$ and would define the escape of the particle, is required. Bearing in mind that the escape occurs when the fluid drag ultimately exceeds the magnetic traction force, we can find, by analysing (19a), the relation

(ii')
$$r_a \cos(\theta - \alpha) < - \sqrt[3]{\left|\frac{V_{ma}}{V_{0a}}\right|}$$

which is the condition used in XSECT in place of (ii).

The action of the XSECT subroutine can be described in terms of values YLA, YL, YHA, YH, which are respectively the initial and current estimates of the lower and upper value of Y, the ordinate of a point on the critical trajectory, and in terms of auxiliary parameters $P$ and $Q$ (see Figure 7).

At the beginning YL = YLA and YH = YHA. The particle is started at the point {XAO, YAO = $\frac{1}{2}$(YLA + YHA)} and the trajectory is produced. If the particle is captured, YL is raised to YAO; if it is not captured, YH is lowered to YAO. The new starting point is set, in either case, as $\frac{1}{2}$(YL + YH). This process is repeated. If the initial estimates have been chosen correctly, i.e. if the unknown Y lies indeed between YLA and YHA, the sequential values YL and YH will converge towards Y. When YL and YH have come sufficiently close to give Y with a required accuracy, the process is halted.

If Y does not lie between YLA and YHA, the starting point would converge upon one of these values. This is, however, avoided by defining the parameters P and Q which lie below YHA and above YLA. Every time the starting point enters the region either between YHA and P or YLA and Q, the values YHA or YLA are redefined until the point Y lies between them. Then Y is found as mentioned previously.

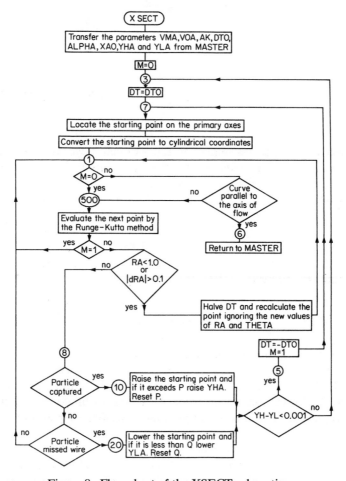

Figure 8. Flow chart of the XSECT subroutine

Having found Y, the time step length is made negative and the trajectory is iterated backwards up to a point C, where the difference between its ordinate YCA and that of the previous iteration is less than a stipulated error. The coordinate YCA is taken as the capture radius, the value XCA gives the distance from the origin to the place where the capture radius has been gauged.

The action of the XSECT subroutine, which has been just described, can be followed from its flow chart shown in Figure 8.

## 4.2 Graph-plotting programs

The curves of the trajectories and velocities produced by the main program require a pictorial representation. Any graph-plotting package can be used for this purpose. As an example we present here two simple graph-plotting programs based on the GINO-F package. Details of the specific commands can be found in the manual *Graph Plotting from Fortran Programs* available from the Computing Laboratory, University of Salford.

The program TRAJECTORIES reads the number of trajectories, the points of the wire contour, and the points of individual trajectories via channels 1, 5, and 6, respectively. The program VELOCITIES reads the number of velocity curves, the velocity range, the angle $\alpha$ via channel 1, and the points of the individual velocity curves via channel 6.

## 4.3 Example

As an example a brief investigation of FeO spherical particles carried by water and captured in the single-wire approximation is presented. The following data, relevant to the problem, have been used in the computation:

| | |
|---|---|
| saturation magnetization of the wire | $M_s = 1.6 \times 10^6 \, \text{Am}^{-1}$ |
| external magnetic field | $H_0 = 1.0 \times 10^6 \, \text{Am}^{-1}$ |
| radius of the wire | $a = 10 \, \mu\text{m}$ |
| radius of the paramagnetic particle | $R = 1 \, \mu\text{m}$ |
| difference between the susceptibilities of the particle and the fluid | $\chi = 7.178 \times 10^{-3}$ |
| viscosity of the fluid | $\eta = 1.0 \times 10^{-3} \, \text{Nm}^{-2} \, \text{s}$ |
| density of the particle | $\rho_p = 5.7 \times 10^3 \, \text{kgm}^{-3}$ |
| magnetic velocity | $V_{ma} = V_m/a = \frac{2}{9}(\chi\mu_0 M_s H_0 R_a^2/\eta)$ $= 3.207 \times 10^4 \, \text{s}^{-1}$ |
| **fluid velocity** | $V_{0a} = -6.288 \times 10^2 \, \text{s}^{-1}$ |

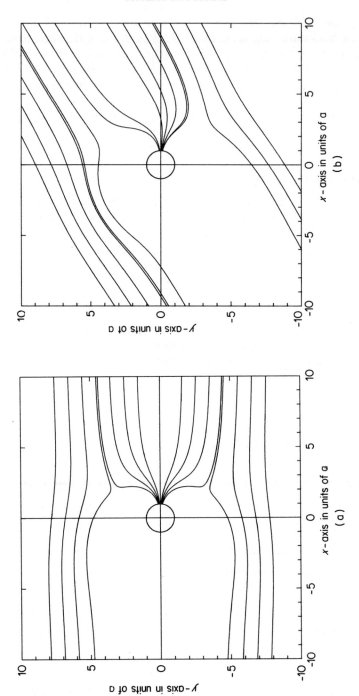

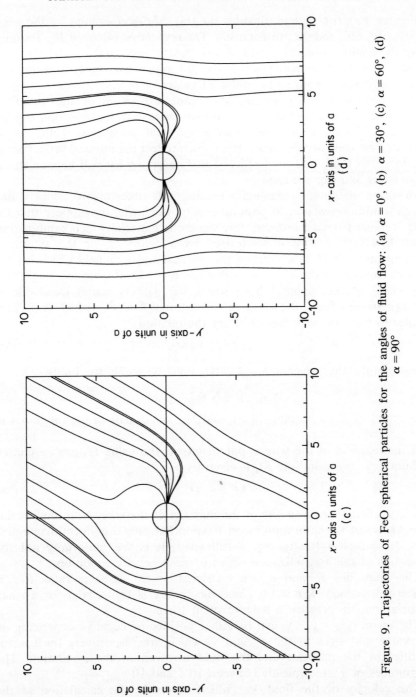

Figure 9. Trajectories of FeO spherical particles for the angles of fluid flow: (a) $\alpha = 0°$, (b) $\alpha = 30°$, (c) $\alpha = 60°$, (d) $\alpha = 90°$

Figures 9(a), (b), (c) and (d) show the trajectories computed for the angle $\alpha = 0°$, 30°, 60°, and 90°, respectively. The respective values of $R_{ca}$ resulting from the computation are 4.613, 4.367, 4.410, and 4.697.

Note that the trajectories which are just super- and subcritical for, say, $\alpha = 0°$ need not remain so for different values of $\alpha$.

It is obvious that by varying the numerical values of the quantities mentioned in this example one can obtain particle trajectories and capture radii for a wide choice of possible physical situations. Thus, as an exercise, one can vary the input data (which reflect the physical properties of the wire, the fluid, and the particles) and observe what will happen to the trajectories and capture radii.

However, to obtain meaningful results, it is necessary to adopt values which would correspond to physical reality. For instance, suppose that the wire is made from an ordinary ferromagnetic material which exhibits only small hysteresis and is characterized by a value $M_s$ of the saturation magnetization. If $H_s$ is a value of the internal magnetic field which is large enough to saturate the wire (this value can be read off from the magnetization curve of the material from which the wire is made) then, due to demagnetizing effects, the value of the external field $H_0$, which is perpendicular to the wire axis, has to satisfy the relation

$$H_0 \geqslant H_s + \tfrac{1}{2}M_s. \tag{36}$$

Consequently, the constant $K = M_s/2H_0$ must lie within the limits

$$0 \leqslant K \leqslant 1. \tag{37}$$

The values of $K$, depending on $M_s$ and $H_0$, are usually in the range 0.4 to 0.9.

If the material of the wire is paramagnetic, the value $H_0$ can be chosen without any constraint and $M_s$ is given by

$$M_s = \chi_m H_0, \tag{38}$$

where $\chi_m$ is the susceptibility of the material of the wire. Note that in this case $M_s$ is not the saturation magnetization but only a uniform magnetization. Nevertheless the theory is still valid; it is the uniformity not the saturation of the magnetization which is the necessary condition.

The quantities $R$ and $a$ are restricted as far as their ratio $R/a$ is concerned, namely $R/a \leqslant 0.1$, since the theory is formulated for a small particle limit. In practice $a$ is larger than 1 $\mu$m.

The quantity $\chi = \chi_p - \chi_f$ can be either positive or negative depending on the value and signs of the $\chi_p$ and $\chi_f$ which are respectively the susceptibilities of the particles and the fluid involved in the separation. The magnitudes of $\chi$ are typically between $10^{-5}$ and $10^{-2}$.

The velocity of the fluid $V_0$ (which enters in the calculation as the normalized velocity $V_{0a} = V_0/a$) can vary over a very broad range, say, from $1 \text{ mm s}^{-1}$ to $50 \text{ mm s}^{-1}$.

To facilitate the legibility of the data and results, the input and output in our programs (sections 4.1 and 4.2) are handled simply in the F-format. The numerical field widths of the format stipulated in the programs should be sufficient to accomodate the results for the data within the indicated range of values. If, however, some results were to require an extension of the field widths, it is easy to change them accordingly or to use the E- instead of the F- format.

## 5. FURTHER EXERCISES

Apart from varying the numerical data, in the manner described in section 4.3 (for instance calculating the trajectories for $\chi < 0$, i.e. for $V_{ma} < 0$), the following, more substantial, exercises are suggested.

(1) Apply the method of solving higher order differential equations (see section 3.3) to equations of particle motion (18a, b), which contain the second-order inertial term. Generalize the main separation program to produce the numerical solutions of these equations.

(2) Consider a capture process, instead of in the single wire, in the single-sphere approximation. Find the magnetic field and the fluid velocity distribution around a ferromagnetic sphere magnetized to saturation by an external field $H_0$ and immersed in fluid flow of a background velocity $V_0$. In analogy with section 2 derive the appropriate equations of particle motion. Modify the main separation program to solve these equations numerically. Compare the capture efficiency of the single wire and the single-sphere approximations.

(3) Analyse a capture process in the single-wire approximation, where the wire has an elliptic cross-section. Similarly, as suggested in section 5.2, find the magnetic field, the fluid velocity distribution, and derive the appropriate equations of particle motion. Modify the FUNCTN sub-routine accordingly and use the main separation program for finding numerical solutions of these equations. Investigate the capture process for various values of eccentricity and various orientations of the elliptic cross-sections in respect to the directions of fluid flow and magnetic field.

## REFERENCES

1. J. H. P. Watson, *J. Appl. Phys.*, **44**, 4209 (1973).
2. R. R. Birss, R. Gerber, and M. R. Parker, *Filtration & Separation*, **July/August,** 1 (1977).

3. R. R. Birss, R. Gerber, and M. R. Parker, *I.E.E.E. Trans. Magn.*, **MAG-12,** 892 (1976).
4. B. I. Bleaney and B. Bleaney, *Electricity and Magnetism*, 2nd edn. (Clarendon Press, Oxford) 47 (1965).
5. W. F. Lawson Jr., W. H. Simons, and R. P. Treat, *J. Appl. Phys.*, **48,** 3213 (1977).
6. D. M. Young and R. T. Gregory, *A Survey of Numerical Mathematics* (Addison-Wesley, Reading, Mass.) 422 (1972).

TYPICAL INPUT DATA FOR THE MAIN PROGRAM (FOR ALPHA = 30 DEGR.)


```
    2    1   16   0.5236
   15.0                 1.6
32071.0      -628.8     0.8      0.0002     0.01
   15.0                 2.6
32071.0      -628.8     0.8      0.0002     0.01
   15.0                 3.6
32071.0      -628.8     0.8      0.0002     0.01
   15.0                 4.48
32071.0      -628.8     0.8      0.0002     0.01
   15.0                 4.6
32071.0      -628.8     0.8      0.0002     0.01
   15.0                 5.6
32071.0      -628.8     0.8      0.0002     0.01
   15.0                 6.6
32071.0      -628.8     0.8      0.0002     0.01
   15.0                 7.6
32071.0      -628.8     0.8      0.0002     0.01
   15.0                -1.6
32071.0      -628.8     0.8      0.0002     0.01
   15.0                -2.6
32071.0      -628.8     0.8      0.0002     0.01
   15.0                -3.6
32071.0      -628.8     0.8      0.0002     0.01
   15.0                -4.48
32071.0      -628.8     0.8      0.0002     0.01
   15.0                -4.6
32071.0      -628.8     0.8      0.0002     0.01
   15.0                -5.6
32071.0      -628.8     0.8      0.0002     0.01
   15.0                -6.6
32071.0      -628.8     0.8      0.0002     0.01
   15.0                -7.6
32071.0      -628.8     0.8      0.0002     0.01
```

TYPICAL OUTPUT FROM THE MAIN PROGRAM (FOR ALPHA = 30 DEGR.)

```
  K=.80    VMA=  32071.00    VOA=   -628.80
VMA/VOA= -51.00    ALPHA=0.523600
  XAO=  15.0000
  YAO=   1.6000
NUMBER OF COMPUTED POINTS  Q= 219
```

| X | Y | V | DRA/DT | DTH/DT |
|---|---|---|---|---|
| 12.19037 | 8.88566 | 628.09008 | -625.39780 | 3.85099 |
| 12.13641 | 8.85351 | 628.08329 | -625.37877 | 3.87576 |
| 12.08246 | 8.82135 | 628.07635 | -625.35959 | 3.90071 |
| 12.02852 | 8.78918 | 628.06925 | -625.34026 | 3.92585 |
| 11.97458 | 8.75701 | 628.06199 | -625.32078 | 3.95118 |
| 11.92065 | 8.72482 | 628.05457 | -625.30115 | 3.97669 |
| 11.86672 | 8.69263 | 628.04698 | -625.28136 | 4.00238 |
| 11.81280 | 8.66042 | 628.03921 | -625.26142 | 4.02827 |
| 11.75889 | 8.62820 | 628.03127 | -625.24132 | 4.05433 |
| 11.70499 | 8.59598 | 628.02315 | -625.22107 | 4.08058 |
| 11.65109 | 8.56374 | 628.01484 | -625.20066 | 4.10702 |
| 11.59720 | 8.53149 | 628.00634 | -625.18009 | 4.13364 |
| 11.54332 | 8.49923 | 627.99764 | -625.15935 | 4.16045 |
| 11.48944 | 8.46697 | 627.98873 | -625.13846 | 4.18744 |
| 11.43558 | 8.43468 | 627.97962 | -625.11740 | 4.21462 |
| 11.38172 | 8.40239 | 627.97030 | -625.09618 | 4.24198 |
| 11.32787 | 8.37009 | 627.96075 | -625.07479 | 4.26952 |
| 11.27403 | 8.33777 | 627.95098 | -625.05324 | 4.29724 |
| 11.22019 | 8.30545 | 627.94097 | -625.03151 | 4.32515 |
| 11.16637 | 8.27311 | 627.93073 | -625.00962 | 4.35323 |
| 11.11255 | 8.24075 | 627.92024 | -624.98757 | 4.38149 |
| 11.05874 | 8.20839 | 627.90950 | -624.96533 | 4.40993 |
| 11.00495 | 8.17601 | 627.89850 | -624.94293 | 4.43854 |
| 10.95116 | 8.14362 | 627.88723 | -624.92036 | 4.46733 |
| 10.89738 | 8.11121 | 627.87569 | -624.89761 | 4.49628 |
| 10.84361 | 8.07879 | 627.86387 | -624.87469 | 4.52541 |
| 10.78985 | 8.04636 | 627.85176 | -624.85159 | 4.55470 |
| 10.73610 | 8.01391 | 627.83935 | -624.82832 | 4.58416 |
| 10.68236 | 7.98145 | 627.82664 | -624.80487 | 4.61378 |
| 10.62863 | 7.94898 | 627.81362 | -624.78124 | 4.64356 |
| 10.57491 | 7.91649 | 627.80028 | -624.75744 | 4.67349 |
| 10.52120 | 7.88398 | 627.78660 | -624.73345 | 4.70358 |
| 10.46750 | 7.85146 | 627.77259 | -624.70929 | 4.73381 |
| 10.41382 | 7.81892 | 627.75822 | -624.68495 | 4.76418 |
| 10.36014 | 7.78636 | 627.74350 | -624.66043 | 4.79470 |
| 10.30648 | 7.75379 | 627.72842 | -624.63573 | 4.82535 |
| 10.25283 | 7.72121 | 627.71295 | -624.61085 | 4.85613 |
| 10.19919 | 7.68860 | 627.69710 | -624.58579 | 4.88703 |
| 10.14556 | 7.65598 | 627.68085 | -624.56055 | 4.91806 |
| 10.09195 | 7.62334 | 627.66418 | -624.53513 | 4.94919 |
| 10.03835 | 7.59068 | 627.64710 | -624.50953 | 4.98043 |
| 9.98476 | 7.55800 | 627.62959 | -624.48375 | 5.01177 |
| 9.93119 | 7.52531 | 627.61164 | -624.45779 | 5.04320 |
| 9.87763 | 7.49259 | 627.59323 | -624.43166 | 5.07472 |

TYPICAL OUTPUT FROM THE MAIN PROGRAM (FOR ALPHA = 30 DEGR.)

| | | | | |
|---|---|---|---|---|
| 9.82409 | 7.45986 | 627.57436 | -624.40535 | 5.10631 |
| 9.77056 | 7.42710 | 627.55500 | -624.37886 | 5.13796 |
| 9.71704 | 7.39433 | 627.53516 | -624.35220 | 5.16968 |
| 9.66354 | 7.36153 | 627.51481 | -624.32537 | 5.20144 |
| 9.61006 | 7.32872 | 627.49394 | -624.29837 | 5.23323 |
| 9.55659 | 7.29588 | 627.47255 | -624.27120 | 5.26506 |
| 9.50313 | 7.26302 | 627.45061 | -624.24386 | 5.29689 |
| 9.44970 | 7.23013 | 627.42811 | -624.21636 | 5.32872 |
| 9.39628 | 7.19722 | 627.40504 | -624.18870 | 5.36055 |
| 9.34288 | 7.16429 | 627.38138 | -624.16088 | 5.39234 |
| 9.28949 | 7.13134 | 627.35713 | -624.13291 | 5.42409 |
| 9.23612 | 7.09836 | 627.33225 | -624.10479 | 5.45578 |
| 9.18278 | 7.06536 | 627.30675 | -624.07652 | 5.48740 |
| 9.12945 | 7.03233 | 627.28060 | -624.04811 | 5.51893 |
| 9.07614 | 6.99927 | 627.25379 | -624.01956 | 5.55034 |

AND SO ON UNTIL THE END OF THE CURVE:

| | | | | |
|---|---|---|---|---|
| 1.99124 | 0.41735 | 5001.67286 | -4677.03717 | -871.26219 |
| 1.93562 | 0.38296 | 5480.27835 | -5160.53217 | -934.83109 |
| 1.87304 | 0.34697 | 6097.52330 | -5784.10413 | -1013.04001 |
| 1.80132 | 0.30904 | 6929.73002 | -6624.48599 | -1112.93695 |
| 1.71689 | 0.26863 | 8125.14141 | -7830.53890 | -1247.62033 |
| 1.66812 | 0.24723 | 8948.72743 | -8660.67437 | -1335.55755 |
| 1.61342 | 0.22484 | 10016.44810 | -9736.00489 | -1444.80580 |
| 1.55086 | 0.20120 | 11465.39899 | -11193.90579 | -1586.01225 |
| 1.47731 | 0.17593 | 13564.56137 | -13303.78436 | -1779.20508 |
| 1.38701 | 0.14839 | 16933.11739 | -16685.52037 | -2068.28761 |
| 1.33225 | 0.13340 | 19551.81429 | -19312.12550 | -2279.52085 |
| 1.26754 | 0.11726 | 23409.29863 | -23178.74971 | -2574.57313 |
| 1.18736 | 0.09943 | 29766.09748 | -29546.41012 | -3029.53802 |
| 1.13808 | 0.08959 | 34867.61021 | -34654.26075 | -3373.57563 |
| 1.07897 | 0.07882 | 42661.96776 | -42455.80182 | -3872.16617 |
| 1.00376 | 0.06665 | 56335.52544 | -56137.70355 | -4688.95543 |

THEN THE SAME OUTPUT FORMAT FOR CURVES NO.:2,3,4,.....,16.

MAIN MAGNETIC SEPARATION PROGRAM


```
C       THE MAIN PROGRAM SERVES AS A CARRIAGE TO INPUT DATA AND OUTPUT RESULTS
C   AFTER THE EQUATIONS HAVE BEEN SOLVED.IT ALSO DIRECTS THE DATA TO THE
C   APPROPRIATE SUBROUTINE THAT IS TO PERFORM THE ITERATIONS ON THE EQUATIONS
C   WHICH ARE HELD IN THE SUBROUTINE 'FUNCTN'.THERE ARE TWO METHODS OF
C   ITERATION,THE RUNGE-KUTTA AND THE ADAMS-MOULTON METHOD.
C   IN ADDITION TO THIS THE SUBROUTINE 'XSECT' WILL FIND THE CAPTURE CROSS-
C   SECTION OF THE WIRE FOR DIFFERENT FLUID VELOCITIES.
C
C       VMA,VOA,AK AND DTO ARE THE MAGNETIC AND FLUID VELOCITIES,
C       THE SHORT RANGE CONSTANT AND THE TIME STEP,RESPECTIVELY.
C       E IS THE MAXIMUM DEVIATION BETWEEN PREDICTOR AND CORRECTOR IN THE
C       PREDICTOR-CORRECTOR METHODS.
C       YLA AND YHA ARE PARAMETERS IN 'XSECT' SUBROUTINE.
C       J IS THE CONTROL INTEGER WHICH SELECTS THE METHOD OF ITERATION.
C       J=1,2,3.
C       K IS THE CONTROL INTEGER WHICH SELECTS THE MODE OF OUTPUT.
C       K=1,2,3.
C       IO IS THE NUMBER OF INITIAL VALUES OF X AND Y,READ IN AS XA AND YA,
C       FOR WHICH CURVES WILL BE PRODUCED.
C       Z IS THE RATIO OF VMA TO VOA.
C       YC IS THE CAPTURE CROSS-SECTION.
C       V IS THE VELOCITY OF THE PARTICLE AT A GIVEN POINT.
C       THE SUBROUTINES WILL PRODUCE Q POINTS(I.E. Q X-Y PAIRS)FOR
C       EACH CURVE.THIS PARAMETER IS IMPORTANT IN THE PLOTTING OF THE
C       CURVES.
C       DAR,DIT AND DEET ARE ARRAYS OF THE GRADIENTS, I.E. THE RADIAL
C       AND ANGULAR VELOCITIES, AND THE TIME STEP,RESPECTIVELY.
C       ALPHA IS THE INCLINATION OF THE FLUID VELOCITY TO THE X-AXIS.
C
      INTEGER Q
      DIMENSION X(1000),Y(1000),G(1000),F(1000),U(1000)
      DIMENSION DAR(1000),DIT(1000),DEET(1000)
      READ(1,1001)J,K,IO,ALPHA
      I=0
  100 I=I+1
      IF(J.EQ.3)GOTO 3
      READ(1,1002)XAO,YAO
      READ(1,1000)VMA,VOA,AK,DTO,E
      XA=XAO*COS(ALPHA)-YAO*SIN(ALPHA)
      YA=YAO*COS(ALPHA)+XAO*SIN(ALPHA)
      ZZ=VMA/VOA
      DT=DTO
      GOTO(1,2),J
    1 CALL RK(XA,YA,VMA,VOA,AK,DT,NN,X,Y,U,ALPHA,DAR,DIT,DEET)
      GOTO 10
    2 CALL AM(XA,YA,VMA,VOA,AK,DT,NN,X,Y,U,E,ALPHA,DAR,DIT,DEET)
      GOTO 10
    3 READ(1,1004)VMA,VOA,AK,DT,YLA,YHA,XAO,ALPHA
      CALL XSECT(YLA,YHA,VMA,VOA,AK,DT,Z,YC,XC,XAO,ALPHA)
      IF(K.EQ.3)GOTO4
      WRITE(2,1006)AK,VMA,VOA,Z,ALPHA,XC,YC
      IF(K.EQ.2)GOTO5
    4 WRITE(3,1005)Z,YC,XC,ALPHA
    5 IF(I.EQ.IO)GOTO 20
      I=I+1
      GOTO 3
```

MAIN MAGNETIC SEPARATION PROGRAM

```
   10 Q=NN
      IF(K.EQ.3)GOTO6
      WRITE(2,1009) AK,VMA,VOA,ZZ,ALPHA,XAO,YAO
      WRITE(2,1007)Q
      WRITE(2,1008)
      WRITE(2,1002)(X(N),Y(N),V(N),DAR(N),DIT(N),DEET(N),N=1,Q)
      WRITE(2,1010)
      IF(K.EQ.2)GOTO7
    6 WRITE(3,1003)Q
      WRITE(3,1002)(X(N),Y(N),V(N),DAR(N),DIT(N),DEET(N),N=1,Q)
    7 IF(I.EQ.IO)GOTO 20
      GOTO 100
   20 STOP
 1000 FORMAT(2F10.2,F10.5,2F10.8,2F10.5)
 1001 FORMAT(3I5,F10.5)
 1002 FORMAT(1X,F10.5,2F20.5,2F15.5,F15.10)
 1003 FORMAT(1X,I10)
 1004 FORMAT(2F10.2,F10.5,F10.8,5F10.5)
 1005 FORMAT(1X,F10.5,3F20.5)
 1006 FORMAT(5H0  K=,F3.2,7H   VMA=,F10.2,
     C7H   VOA=,F10.2/10H  VMA/VOA=,
     CF7.2,9H   ALPHA=,F8.6/7H   XCA=,
     CF9.4/7H   RCA=,F9.4)
 1007 FORMAT(1X,31H  NUMBER OF COMPUTED POINTS  Q=,I4///)
 1008 FORMAT(1X,88H    X                Y                 U
     C   DRA/DT        DTH/DT       DT///)
 1009 FORMAT(5H0  K=,F3.2,7H   VMA=,F10.2,
     C7H   VOA=,F10.2/10H  VMA/VOA=,
     CF7.2,9H   ALPHA=,F8.6/7H   XAO=,
     CF9.4/7H   YAO=,F9.4)
 1010 FORMAT(////)
      END
C
C
C   THE FUNCTN SUBROUTINE.
C
C   THIS SUBROUTINE CONTAINS THE EQUATIONS TO BE ITERATED.
C
      SUBROUTINE FUNCTN(RA,THETA,VMA,VOA,AK,DRADT,DTHDT,ALPHA)
      DRADT=VOA*(1.0-(RA**(-2)))*COS(THETA-ALPHA)-VMA*AK/(RA**5)-VMA*
     1COS(2.0*THETA)/(RA**3)
      DTHDT=-VOA*(1.0+(RA**(-2)))*SIN(THETA-ALPHA)/RA-VMA*SIN(2.0*THETA
     1)/(RA**4)
      RETURN
      END
C
C
C   THE RUNGE-KUTTA METHOD.
C
C   THIS IS A SELF-STARTING ,FOURTH-ORDER METHOD OF HIGH ACCURACY.
C   IT INCORPORATES THE STEP LENGTH ADJUSTMENT.
C
      SUBROUTINE RK(XA,YA,VMA,VOA,AK,DT,N,X,Y,U,ALPHA,DAR,DIT,DEET)
      REAL L,KR1,KT1,KR2,KT2,KR3,KT3,KR4,KT4
      DIMENSION X(1000),Y(1000),V(1000)
      DIMENSION DAR(1000),DIT(1000),DEET(1000)
```

```
      N=1
      X(N)=XA
      Y(N)=YA
      RA=SQRT(XA*XA+YA*YA)
      THETA=ATAN2(YA,XA)
      Z=RA+0.2
      IF(Z.LE.14.2)Z=14.2
    1 R=RA
      TH=THETA
   40 CALL FUNCTN(RA,THETA,VMA,VOA,AK,DRADT,DTHDT,ALPHA)
      V(N)=SQRT(DRADT*DRADT+RA*RA*DTHDT*DTHDT)
      DAR(N)=DRADT
      DIT(N)=DTHDT
      DEET(N)=DT
      IF(N.EQ.1000)GOTO 2
      IF(RA.LT.1.01.AND.RA.GT.1.00)GOTO 2
      IF(RA.GT.Z)GOTO 2
      KR1=DRADT*DT
      KT1=DTHDT*DT
      RA=R+0.5*KR1
      THETA=TH+0.5*KT1
      CALL FUNCTN(RA,THETA,VMA,VOA,AK,DRADT,DTHDT,ALPHA)
      KR2=DRADT*DT
      KT2=DTHDT*DT
      RA=R+0.5*KR2
      THETA=TH+0.5*KT2
      CALL FUNCTN(RA,THETA,VMA,VOA,AK,DRADT,DTHDT,ALPHA)
      KR3=DRADT*DT
      KT3=DTHDT*DT
      RA=R+KR3
      THETA=TH+KT3
      CALL FUNCTN(RA,THETA,VMA,VOA,AK,DRADT,DTHDT,ALPHA)
      KR4=DRADT*DT
      KT4=DTHDT*DT
      DRA=(KR1+2.0*(KR2+KR3)+KR4)/6.0
      DTH=(KT1+2.0*(KT2+KT3)+KT4)/6.0
      RA=R+DRA
      THETA=TH+DTH
      IF(RA.LT.1.00.OR.ABS(DRA).GT.0.1)GOTO 4
      N=N+1
      X(N)=RA*COS(THETA)
      Y(N)=RA*SIN(THETA)
      GOTO 1
    4 DT=DT/2.0
      RA=R
      THETA=TH
      GOTO 40
    2 RETURN
      END
C
C
C   THE ADAMS-MOULTON METHOD.
C
C   THIS METHOD USES THE FIFTH-ORDER ADAMS-MOULTON PREDICTOR-
C   CORRECTOR METHOD UTILISING THE RUNGE-KUTTA METHOD AS ITS STARTER
C   TO FIND THE FIRST FOUR POINTS.
```

MAIN MAGNETIC SEPARATION PROGRAM

```
C   THE ERROR IN THIS TECHNIQUE IS GOVERNED BY E.
C   IT ALSO INCORPORATES THE STEP LENGTH ADJUSTMENT
C
        SUBROUTINE AM(XA,YA,VMA,VOA,AK,DT,N,X,Y,V,E,ALPHA,F,G,DEET)
        REAL L,KR1,KT1,KR2,KT2,KR3,KT3,KR4,KT4
        DIMENSION X(1000),Y(1000),F(1000),G(1000),V(1000)
        DIMENSION DEET(1000)
        NO=4
        N=1
        X(N)=XA
        Y(N)=YA
        RA=SQRT(XA*XA+YA*YA)
        THETA=ATAN2(YA,XA)
        Z=RA+0.2
        IF(Z.LE.14.2)Z=14.2
C   START OF RUNGE-KUTTA
      1 R=RA
        TH=THETA
     40 CALL FUNCTN(RA,THETA,VMA,VOA,AK,DRADT,DTHDT,ALPHA)
        V(N)=SQRT(DRADT*DRADT+RA*RA*DTHDT*DTHDT)
        DEET(N)=DT
        F(N)=DRADT
        G(N)=DTHDT
        IF(N.EQ.1000)GOTO 3
        IF(RA.LE.1.01.AND.RA.GE.1.00)GOTO 3
        IF(RA.GE.Z)GOTO 3
        IF(N.GE.NO)GOTO 2
        KR1=DRADT*DT
        KT1=DTHDT*DT
        RA=R+0.5*KR1
        THETA=TH+0.5*KT1
        CALL FUNCTN(RA,THETA,VMA,VOA,AK,DRADT,DTHDT,ALPHA)
        KR2=DRADT*DT
        KT2=DTHDT*DT
        RA=R+0.5*KR2
        THETA=TH+0.5*KT2
        CALL FUNCTN(RA,THETA,VMA,VOA,AK,DRADT,DTHDT,ALPHA)
        KR3=DRADT*DT
        KT3=DTHDT*DT
        RA=R+KR3
        THETA=TH+KT3
        CALL FUNCTN(RA,THETA,VMA,VOA,AK,DRADT,DTHDT,ALPHA)
        KR4=DRADT*DT
        KT4=DTHDT*DT
        DRA=(KR1+2.0*(KR2+KR3)+KR4)/6.0
        DTH=(KT1+2.0*(KT2+KT3)+KT4)/6.0
        RA=R+DRA
        THETA=TH+DTH
        IF(RA.LT.1.00.OR.ABS(DRA).GT.0.1)GOTO 4
        N=N+1
        X(N)=RA*COS(THETA)
        Y(N)=RA*SIN(THETA)
        GOTO 1
      4 DT=DT/2.0
        NO=N+4
        RA=R
```

MAIN MAGNETIC SEPARATION PROGRAM

```
       THETA=TH
       GOTO 40
C    START OF ADAMS-MOULTON
    2 R=RA
       TH=THETA
       F1=F(N)
       G1=G(N)
       N=N-1
       F2=F(N)
       G2=G(N)
       N=N-1
       F3=F(N)
       G3=G(N)
       N=N-1
       F4=F(N)
       G4=G(N)
       N=N+3
       AMR1=R+DT*(55.0*F1-59.0*F2+37.0*F3-9.0*F4)/24.0
       AMT1=TH+DT*(55.0*G1-59.0*G2+37.0*G3-9.0*G4)/24.0
       RA=AMR1
       THETA=AMT1
       CALL FUNCTN(RA,THETA,VMA,VOA,AK,DRADT,DTHDT,ALPHA)
       AMR2=R+DT*(9.0*DRADT+19.0*F1-5.0*F2+F3)/24.0
       AMT2=TH+DT*(9.0*DTHDT+19.0*G1-5.0*G2+G3)/24.0
       ETA=ABS(AMR1-AMR2)/AMR2
       IF(ETA.GE.E)GOTO 600
       IF(AMT2.LT.1.0E-10)GOTO 700
       ZETA=ABS(AMT1-AMT2)/AMT2
       IF(ZETA.GE.E)GOTO 600
       GOTO 700
  600 DT=DT/2.0
       NO=N+4
       RA=R
       THETA=TH
       GOTO 1
  700 RA=AMR2
       THETA=AMT2
       IF(RA.LT.1.00.OR.ABS(RA-R).GT.0.1)GOTO 600
       CALL FUNCTN(RA,THETA,VMA,VOA,AK,DRADT,DTHDT,ALPHA)
       N=N+1
       F(N)=DRADT
       G(N)=DTHDT
       V(N)=SQRT(DRADT*DRADT+RA*RA*DTHDT*DTHDT)
       DEET(N)=DT
       X(N)=RA*COS(THETA)
       Y(N)=RA*SIN(THETA)
       IF(N.EQ.1000)GOTO 3
       IF(RA.LE.1.01.AND.RA.GE.1.00)GOTO 3
       IF(RA.GT.2)GOTO 3
       GOTO 2
    3 RETURN
       END
C
C
C    THE XSECT SUBROUTINE.
C
```

MAIN MAGNETIC SEPARATION PROGRAM

```
C   THIS SUBROUTINE CALCULATES THE CAPTURE CROSS SECTION OF THE WIRE.
C   IT DOES SO BY ITERATING A CURVE AND ASSESING WHETHER
C   CAPTURE HAS TAKEN PLACE OR NOT.IF IT HAS, THE STARTING POINT WAS
C   TOO LOW AND IT IS THEN RAISED TO HALF-WAY BETWEEN ITS PRESENT
C   POSITION AND THE STARTING VALUE OF THE CURVE WHEN THE PARTICLE LAST
C   MISSED THE WIRE WHICH IS A LITTLE TOO HIGH.SIMILARLY THE HIGH VALUE
C   IS LOWERED IF THE PARTICLE MISSES THE WIRE.BY REPEATING THIS
C   PROCESS UNTIL THE HIGH AND LOW STARTING POINTS CONVERGE THE CAPTURE
C   CROSS-SECTION IS FOUND.THIS IS REPEATED AS  A FUNCTION OF THE
C   RATIO VMA:VOA.
C
      SUBROUTINE XSECT(YLA,YHA,VMA,VOA,AK,DTO,Z,YC,XC,XAO,ALPHA)
      REAL L,KR1,KT1,KR2,KT2,KR3,KT3,KR4,KT4
      WRITE(2,1040)
 1040 FORMAT(13H0  ITERATIONS//)
      M=0
      D=0.0
      Z=VMA/VOA
      YAO=(YLA+YHA)/2.0
      YH=YHA
      YL=YLA
    3 DT=DTO
    7 YA=XAO*SIN(ALPHA)+YAO*COS(ALPHA)
      XA=XAO*COS(ALPHA)-YAO*SIN(ALPHA)
      RA=SQRT(XA*XA+YA*YA)
      THETA=ATAN2(YA,XA)
    1 R=RA
      TH=THETA
      CALL FUNCTN(RA,THETA,VMA,VOA,AK,DRADT,DTHDT,ALPHA)
      IF(M.EQ.0)GOTO 500
      YC=RA*SIN(THETA-ALPHA)
      XC=RA*COS(THETA-ALPHA)
      DRC=ABS(YC-D)
      IF(DRC.LT.0.0005)GOTO 6
      D=YC
  500 KR1=DRADT*DT
      KT1=DTHDT*DT
      RA=R+0.5*KR1
      THETA=TH+0.5*KT1
      CALL FUNCTN(RA,THETA,VMA,VOA,AK,DRADT,DTHDT,ALPHA)
      KR2=DRADT*DT
      KT2=DTHDT*DT
      RA=R+0.5*KR2
      THETA=TH+0.5*KT2
      CALL FUNCTN(RA,THETA,VMA,VOA,AK,DRADT,DTHDT,ALPHA)
      KR3=DRADT*DT
      KT3=DTHDT*DT
      RA=R+KR3
      THETA=TH+KT3
      CALL FUNCTN(RA,THETA,VMA,VOA,AK,DRADT,DTHDT,ALPHA)
      KR4=DRADT*DT
      KT4=DTHDT*DT
      DRA=(KR1+2.0*(KR2+KR3)+KR4)/6.0
      DTH=(KT1+2.0*(KT2+KT3)+KT4)/6.0
      RA=R+DRA
      THETA=TH+DTH
```

MAIN MAGNETIC SEPARATION PROGRAM

```
        IF(M.EQ.1)GOTO 1
        IF(RA.GT.1.00.AND.ABS(DRA).LT.0.1)GOTO 8
        RA=R
        THETA=TH
        DT=DT/2.0
        GOTO 1
     8  IF(RA.GT.1.00.AND.RA.LT.1.01)GOTO 10
        RL=-(ABS(Z)**0.3333)
        IF(RA*COS(THETA-ALPHA).LT.RL)GOTO 20
        GOTO 1
    10  YL=YAO
        P=YHA-0.1
        WRITE(2,1020)YAO
  1020  FORMAT(1X,F20.5)
        IF(YL.LT.P)GOTO 100
        YHA=YHA+0.2
        YH=YHA
   100  YAO=(YH+YL)/2.0
        W=YH-YL
        IF(W.LT.0.001)GOTO 5
        GOTO 3
    20  YH=YAO
        Q=YLA+0.1
        WRITE(2,1030)YAO
  1030  FORMAT(1X,F30.5)
        IF(YH.GT.Q)GOTO 200
        YLA=YLA/2.0
        IF(YLA.LT.0.05)YLA=0.0
        YL=YLA
   200  YAO=(YH+YL)/2.0
        W=YH-YL
        IF(W.LT.0.001)GOTO 5
        GOTO 3
     5  DT=-DTO
        M=1
        WRITE(2,1010)YAO
  1010  FORMAT(1X,F10.5//)
        GOTO 7
     6  RETURN
        END
```

GRAPH PLOTTING PROGRAM: TRAJECTORIES

```
C  THIS PROGRAM WILL PLOT OUT THE WIRE AND THE PATHS OF THE PARTICLES.
C  THE POINTS FOR THE WIRE SHOULD BE READ IN VIA CHANNEL NO.5.
C  THE DATA FOR THE PATHS SHOULD BE READ IN VIA CHANNEL NO.6.
C
      INTEGER Q
      DIMENSION X(1000),Y(1000),IPLOT(19)
      DATA IPLOT/22,22H PARTICLE TRAJECTORIES,20,20HX-AXIS IN UNITS OF A
     *,20,20HY-AXIS IN UNITS OF A/
      I=0
      READ(1,1001)IO
      READ(5,1000)(X(N),Y(N),N=1,120)
      CALL NARROW
      CALL RSIZE(150.0,150.0)
      CALL FIXAXS(-9.99,10.0,-9.99,10.0)
      CALL FGPLT(X,Y,120,5,0,1,0,IPLOT)
    1 I=I+1
      READ(6,999)Q
      READ(6,1000)(X(N),Y(N),N=1,Q)
      CALL FGPLT(X,Y,Q,15,0,1,1,IPLOT)
      IF(I.EQ.IO)GOTO 100
      GOTO 1
  100 CALL DEVFIN
      STOP
  999 FORMAT(I10)
 1000 FORMAT(1X,F10.5,F20.5)
 1001 FORMAT(I5)
      END
```

GRAPH PLOTTING PROGRAM: VELOCITIES

```
C   THIS PROGRAM WILL PLOT OUT THE LINE REPRESENTING THE
C   VELOCITIES OF THE PARTICLES AS A FUNCTION OF X-DISTANCE FROM
C   THE WIRE,BUT NOT THE WIRE ITSELF AS THE SCALE IS TOO LARGE.
C   THE POINTS FOR THE CURVE SHOULD BE READ IN VIA CHANNEL NO.6.
C
      INTEGER Q
      DIMENSION X(1000),Y(1000),R(1000),V(1000),IPLOT(22)
      DATA IPLOT/19,19HPARTICLE VELOCITIES,22,22HPOSITION IN UNITS OF A,
     *31,31HVELOCITY IN UNITS OF A PER SEC./
      I=0
      READ(1,1001)IO,Z,ALPHA
      READ(6,999)Q
      N=1
    2 READ(6,1000)X(N),Y(N),V(N)
      THETA=ATAN2(Y(N),X(N))
      R(N)=SQRT(X(N)*X(N)+Y(N)*Y(N))*COS(THETA-ALPHA)
      N=N+1
      IF(N.LE.Q)GOTO 2
      CALL NARROW
      CALL RSIZE(200.0,100.0)
      CALL FIXAXS(-9.99,10.0,0.0,Z)
      CALL CHAPEN(2)
      CALL FGPLT(R,V,Q,15,0,1,0,IPLOT)
      I=I+1
      IF(IO.EQ.I)GOTO 1
      DO 1 I=2,IO
      READ(6,999)Q
      N=1
    3 READ(6,1000)X(N),Y(N),V(N)
      THETA=ATAN2(Y(N),X(N))
      R(N)=SQRT(X(N)*X(N)+Y(N)*Y(N))*COS(THETA-ALPHA)
      N=N+1
      IF(N.LE.Q)GOTO 3
      CALL CHAPEN(2)
      CALL FGPLT(R,V,Q,15,0,1,1,IPLOT)
    1 CONTINUE
      CALL DEVFIN
      STOP
  999 FORMAT(I10)
 1000 FORMAT(1X,F10.5,2F20.5)
 1001 FORMAT(I5,2F15.5)
      END
```

Physics Programs
Edited by A. D. Boardman
© 1980 John Wiley & Sons Ltd.

## CHAPTER 6

# Magnetization in the Crystal Field System Praseodymium

J. A. G. TEMPLE

## 1. INTRODUCTION

The rare earth metals, atomic numbers 58–71, have unfilled 4f shells lying inside closed 5s 5p shells. The outermost electrons, that is $5d^1 6s^2$, are lost to the conduction band and become a part of the nearly free electron sea. The small spatial extent of the 4f wavefunctions makes overlap of these wave functions on neighbouring ions very unlikely in the solid. This implies a magnetic moment which is well localized on the ionic site. This situation is very different from the more conventional magnetic systems (Fe, Co, Ni, etc.), in which the magnetism is due entirely to the free electrons, and cannot be attributed to well-localized magnetic moments. For this reason the rare earths are intrinsically interesting, especially so when the range and type of magnetic orderings (ferromagnetic, antiferromagnetic, helical, fan, and spiral types often with magnetic unit cells many times larger than the crystal unit cell) are noted. They are becoming increasingly important in technology too, with uses in high-power permanent magnets and in lasers.

This localized magnetic moment is, to some extent, screened by the 5s 5p electrons and does not experience the full effect of the Coulomb interaction due to the periodic array of ions. Another way of describing this is to say that the crystalline electric field is weaker. Each of the localized magnetic moments experiences the effect of many others in the crystal through a double interaction with the conduction electrons. A local moment will scatter a conduction electron with a spin-dependent interaction so that the conduction electron carries away with it a 'memory' of the orientation of the first moment. This electron, on a second scattering, with a, presumably, different local moment, conveys this memory to the second moment. Thus it appears as though there is a direct interaction between local moments $i$ and $j$ of the form $F(\mathbf{R}_{ij})(\mathbf{J}_i \cdot \mathbf{J}_j)$. $F(\mathbf{R}_{ij})$ oscillates rapidly with the distance $\mathbf{R}_{ij}$

between total ionic angular momenta $J_i$ and $J_j$. This is the well-known RKKY interaction (Rudermann and Kittel,[1] Kasuya,[2] Yosida[3]) which dominates the magnetic behaviour of the 14 rare earth elements. In its simplest form, this interaction may be represented by an internal local magnetic field at each site.

Hund's rules give the ground state of the 4f configuration. This is separated from the other multiplets of total angular momentum by an energy which is typically of the order of 1000 K. At low temperatures (i.e. a few kelvin) therefore, only the lowest $J$-multiplet is occupied; $J$ being the total angular momentum of the ion. This multiplet is $(2J+1)$-fold degenerate in the absence of an internal magnetic field and a crystalline electric field. In the presence of a crystal field, the multiplet is split into a series of singlets, doublets and triplets depending on the symmetry of the charge distributions of neighbouring ions.

At low temperatures (i.e. tens of kelvin) when only a few of these singlets, doublets, and triplets are occupied the system will, in general, be magnetically ordered with all the localized moments arranged in some well-defined pattern. The free electrons, that is those responsible for the electrical conductivity, will sense this ordering and will be scattered differently in the ordered state to the disordered state. There will thus be, possibly, interesting anomalies in the conductivity, or its reciprocal the resistivity, at low temperatures, as well as the magnetization itself. These anomalies reflect the competition between the internal magnetic fields at each lattice site and the local electrical fields due to the Coulomb interaction with neighbouring ions. For praseodymium this competition is finely balanced and, as explained in section 2, no other metal is known to sit so close to a threshold between the ability to order spontaneously as the temperature is reduced and the ability to remain paramagnetic even towards the absolute zero of temperature. The magnetization anomalies may thus be expected to be rather interesting in the presence of an externally applied magnetic field.

## 2. THE MODEL

Praseodymium (Pr) is element number 59 and has therefore a $4f^2$ configuration for which Hund's rules give the quantum numbers $S = 1$, $L = 5$ and, since the shell is less than half full, $J = L - S = 4$. The ground state is denoted $^3H_4$ and is ninefold degenerate in the absence of a crystalline electric field or internal magnetic field. Other ways of adding together the total orbital angular momentum $L$ and the total spin $S$ would give the configurations $^3H_5^3H_6$. These lie at much greater energies; in praseodymium it requires at least 1000 K to excite the ions into these higher energy states. Because of this large energy requirement it is safe to assume only the lowest multiplet, namely $^3H_4$, is populated at room temperature or below.

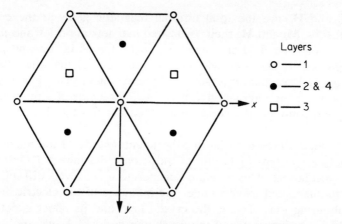

Figure 1. Plan view of dhcp crystal structure showing layers
of ions in the planes perpendicular to the crystal $c$-axis. Ions
in successive layers are denoted by 1, 2, 3, and 4

The solid-state form of Pr has the double hexagonal close-packed structure (dhcp) as shown in Figure 1. This is usually denoted as a stacking
sequence ABCB of hexagonal layers. In a material with cubic symmetry the
layer sequence is ABC and in a hexagonal material the sequence is ABAB.
It is clear that, based on the nearest-neighbour environments, the dhcp
structure has layers of alternate designation, cubic or hexagonal. Two types
of crystallographic symmetry imply that there are two different magnetic
behaviours; in this material two interpenetrating sublattices each with its
own molecular field.

Suppose that there is an externally applied magnetic field. This acts at
each lattice site, together with an additional magnetic field due to an extra
alignment of all the magnetic moments on the neighbouring lattice sites,
since each ion responds to the applied field by a tendency to align with it.
This additional internal field is called a 'molecular field' after the original
concept of Weiss.[4] In this approximation, the magnetic field due to the
partial alignment of neighbouring moments is linearly related to this alignment or magnetization. Although the applied field will be the same at all
lattice sites, the internal field will vary according to the physical arrangement
of the neighbours, that is to the symmetry and number of neighbouring ions
and their magnetization, i.e.

$$\mathbf{H}_c = \frac{\lambda}{g^2 \mu_B^2} \mathbf{M}_c + \frac{\mu}{g^2 \mu_B^2} \mathbf{M}_h + \mathbf{H}_{applied},$$

$$\mathbf{H}_h = \frac{\lambda}{g^2 \mu_B^2} \mathbf{M}_h + \frac{\mu}{g^2 \mu_B^2} \mathbf{M}_c + \mathbf{H}_{applied},$$

$$(1)$$

where $H_c$ and $H_h$ are the total internal magnetic fields at the cubic and hexagonal sites, $M_c$ and $M_h$ their associated magnetizations, $\lambda$ and $\mu$ are the two Weiss molecular field parameters, $g$ is the Landé factor, and $\mu_B$ is the Bohr magneton. It has been assumed in equation (1) that the cubic–cubic and hexagonal–hexagonal interactions are equivalent. This is a reasonable assumption since the interaction between two moments of a pure rare earth metal depends only on the conduction electrons and the distance between the ions.

As the temperatures used in an experimental study of the magnetization will be only a few tens of Kelvin at most, only the lowest $J$-multiplet will enter the calculations. This ninefold degenerate set of levels will split up into singlets, doublets, and triplets under the influence of the electric field of all the neighbouring ions. This is the crystal field and its effect is determined principally by the symmetry of the neighbouring ions. As we are concerned only with a set of levels, all of which possess the same total angular momentum $J$, it is possible to calculate the effect of the Coulomb interaction of all the ions on each other; a kind of Stark effect, by using the method of Stevens operators. These make easier the task of determining the splitting of the levels with the same $J$ but differing $J_z$ through the first-order perturbation caused by the electric field of neighbouring ions. Using these Stevens operator equivalents (Stevens[5]), the ionic Hamiltonian, or total energy operator, becomes

$$\mathcal{H}_\alpha = V_\alpha - g\mu_B \mathbf{H} \cdot \mathbf{J}, \qquad (2)$$

where $\alpha = $ cubic or hexagonal, $V_\alpha$ is the energy in the crystal field, and $\mathbf{H}$ is the total internal magnetic field.

$$V_{cubic} = B_2 O_2^0 + B_4 (O_4^0 + 20\sqrt{2} O_4^3) + B_6 \left( O_6^0 - \frac{35}{\sqrt{8}} O_6^3 + \frac{77}{8} O_6^6 \right), \qquad (3)$$

$$V_{hexagonal} = B_2 O_2^0 + B_4 O_4^0 + B_6 (O_6^0 + \tfrac{77}{8} O_6^6), \qquad (4)$$

where $B_l$ are constants with dimensions of energy and the $O_l^m$ are the Stevens operators that are polynomials of degree $l$ in the quantum mechanical operators $J_z$, $J+$, $J-$. Some typical Stevens operators are:

$$\begin{aligned} O_2^0 &= 3J_z^2 - J(J+1), \\ O_6^6 &= \tfrac{1}{2}\{(J+)^6 + (J-)^6\}, \end{aligned} \qquad (5)$$

and a full list of them may be found in Hutchings.[6] The resulting operator $\mathcal{H}_\alpha$ may be expressed as a matrix whose elements are the numbers obtained from $\langle i | \mathcal{H}_\alpha | j \rangle$, where $|i\rangle$ and $|j\rangle$ are states of total angular momentum $J$ and $z$-component $J_z$. That is, a state of the system is written $|J; J_z\rangle$ where $L$ and

$S$ have been taken as fixed ($L = 5$, $S = 1$ for two 4f electrons). The magnitude of $J = \sqrt{J(J+1)}\hbar$ and the basis set of states is taken to be

$$|J; J_z = J\rangle, \qquad |J; J_z = J - 1\rangle, \ldots, \qquad |J; J_z = -J\rangle.$$

As we are dealing only with the lowest $J$ multiplet, this notation is further abbreviated to $|i\rangle$, $|j\rangle$, etc. where only the $J_z$ component of the total angular momentum is shown within the Bra-Ket notation of Dirac. If the resulting matrix is diagonalized then the eigenvalues displayed down the diagonal are the energies of the different crystal field states in the total internal magnetic field. The corresponding eigenvectors are the wave functions of the state produced and will, in general, be linear combinations of the basis states. In the absence of any internal magnetic field, the crystal field lifts the nine-fold degeneracy of the $^3H_4$ ground state. However, there still remains a partial degeneracy, for example $|+1\rangle$ will have the same energy as $|-1\rangle$ and this degeneracy will be removed by an internal magnetic field. Also, the crystal field reorders the energy levels so that, for example, $|0\rangle$ lies below both $|+1\rangle$ and $|-1\rangle$. When a magnetic field develops due to either an applied magnetic field or spontaneous ordering, this level scheme will change and in the limit of very large internal magnetic fields the levels will be, in ascending order: $|-4\rangle$, $|-3\rangle$, $|-2\rangle$, $|-1\rangle$, $|0\rangle$, $|1\rangle$, $|2\rangle$, $|3\rangle$, $|4\rangle$.

The ground state, which is the state of lowest energy, may be a pure $|J_z\rangle = |0\rangle$ state and is then an example of a 'singlet ground state system'. A characteristic of such a system is that it will not order even at the lowest obtainable temperature because the magnetic moment operator vanishes identically (i.e. $\langle 0 |J_z| 0 \rangle = 0$). Such a system may exist if the ratio of crystal field energy to internal magnetic energy exceeds a certain threshold value. Praseodymium is a unique metal because the ratio of exchange energy (the magnetic interaction between two local moments), to the energy of the ions in the crystal field is found to be about 0.96, whereas the threshold value is 1.0. For a ratio greater than 1.0 Pr would order spontaneously at some small non-zero temperature, but for ratios less than 1.0 it will remain paramagnetic until temperatures of a few millikelvin at which the hyperfine interaction becomes important. No other metal is known to sit so close to this threshold value. The level schemes, from neutron-scattering experiments, in dhcp Pr are shown in Figure 2. From this figure it will be observed that the cubic sites may be expected to play a rather small role in the magnetization at low temperatures since no significant population of levels with resultant moments can occur until temperatures of about 80 K are reached. It is, therefore, instructive to look at the bottom three levels on the hexagonal sites and investigate their behaviour in magnetic fields applied parallel to the $z$- or $x$-axes. This will give physical insight into the behaviour of the system as a whole.

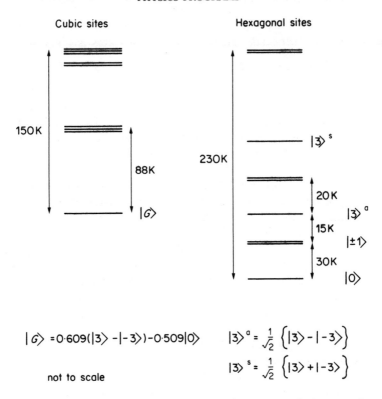

Figure 2. Praseodymium: energy levels in zero applied magnetic
field

## 2.1 Example of way model is used

Ignoring, for the present, any molecular field, we may consider the artificial
system of a $J = 1$ spin sitting in a crystal field of axial symmetry. If instead of
$V_\alpha$ we write $\mathscr{H}_{cf}$ for the effect of the crystalline electric field, then equation
(2) becomes

$$\mathscr{H} = \mathscr{H}_{cf} - g\mu_B \mathbf{H} \cdot \mathbf{J}, \qquad (6)$$

where $\mathbf{H}$ is the applied magnetic field only. Considering first the field applied
parallel to the $z$-axis, then $\mathbf{H} \cdot \mathbf{J} = H_z J_z$. Hence, if we write $\alpha = g\mu_B$ and the
familiar result for $\mathscr{H}_{cf}$ we may express (6) as

$$\mathscr{H} = D J_z^2 - \alpha J_z, \qquad (7)$$

where $D$ and $\alpha$ are numbers with dimensions of energy and $D$ represents
the crystal field splitting in a system of uniaxial symmetry. For a $J = 1$ system

the corresponding matrix Hamiltonian becomes (see for example Dicke and Wittke[7])

$$\mathcal{H} = \begin{pmatrix} D - \alpha & 0 & 0 \\ 0 & 0 & 0 \\ 0 & 0 & D + \alpha \end{pmatrix},$$

where we have used a quantization scheme along the $z$-axis with eigenvalues $0$, $D - \alpha$, $D + \alpha$, so that the energy level diagram has the form shown in Figure 3.

The magnetization is calculated from

$$M = -g\mu_B\langle\langle J_z \rangle\rangle, \tag{8}$$

where $\langle\langle A \rangle\rangle$ denotes the thermal average of the expectation value of the operator $A$. It is necessary to use the thermal average as the energy splitting of the levels caused by either crystal field or magnetic field is of the same order as the temperature and therefore there will be a statistical population of the levels throughout the $10^{23}$ sites in a macroscopic crystal.

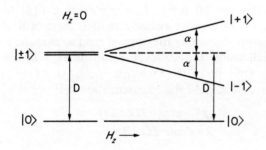

Figure 3. Energy levels of a $J = 1$ system with uniaxial crystal field showing how the levels split with a magnetic field applied along the $z$-axis.

With the eigenvalues and eigenvectors given in Figure 3 we have

$$\langle\langle J_z \rangle\rangle = \left[ \langle 0| J_z |0\rangle + \langle -1| J_z |-1\rangle \exp\left\{-\frac{(D-\alpha)}{kT}\right\} \right.$$

$$\left. + \langle +1| J_z |+1\rangle \exp\{-(D+\alpha)/kT\} \right] Z^{-1}, \qquad (9)$$

where $Z$, the partition function, is given by

$$Z = 1 + \exp\{-(D-\alpha)/kT\} + \exp\{-(D+\alpha)/kT\}, \qquad (10)$$

$k$ is Boltzmann's constant and $T$ is the temperature in kelvin. Clearly, when $\alpha = 0$ ($H_z = 0$) then $\langle\langle J_z \rangle\rangle = 0$ and the system can never order no matter how low the temperature to which it is subjected. This is characteristic of a singlet ground state system. However, when $\alpha \neq 0$ (since $\langle 0| J_z |0\rangle = 0$) the magnetization looks like

$$M = -\frac{g\mu_B}{Z}[-\exp\{-(D-\alpha)/kT\} + \exp\{-(D+\alpha)/kT\}], \qquad (11)$$

This is very small for $\alpha \ll D$, (i.e. energy of the spin in the applied magnetic field is very much less than the crystal field splitting), showing that the initial susceptibility is small. If the applied field is increased, then, eventually, the $|-1\rangle$ level will fall below the $|0\rangle$ level. This occurs at a field for which $D - \alpha = 0$ or $H_z = D/g\mu_B$. In a more realistic model, there will also be a molecular field which tends to amplify the applied field so that the condition for the $|-1\rangle$ level to fall below the $|0\rangle$ level would be $H_z + \beta M_z = D/g\mu_B$, where $\beta$ represents the strength of the molecular field. At a field infinitesimally greater than this critical field the magnetization will become

$$M = \frac{g\mu_B[1 - \exp(-2D/kT)]}{2 + \exp(-2D/kT)}, \qquad (12)$$

which for very small temperatures tends to $M = (\frac{1}{2})g\mu_B$. This appears as a jump discontinuity in the magnetization, i.e. for $\alpha$ just smaller than $D$, as the temperature tends to zero, the magnetization remains zero; but for $\alpha$ just larger than $D$, under the same circumstances, the magnetization is $(\frac{1}{2})g\mu_B$.

Now consider what happens if the field is applied along the $x$-direction, in the simple model. If we put $\alpha = g\mu_B H_x$ then

$$\mathcal{H} = DJ_z^2 - \alpha J_x, \qquad (13)$$

or in matrix form, still using quantization along the $z$-axis,

$$\mathcal{H} = \begin{pmatrix} D & 0 & 0 \\ 0 & 0 & 0 \\ 0 & 0 & D \end{pmatrix} - \frac{\alpha}{\sqrt{2}} \begin{pmatrix} 0 & 1 & 0 \\ 1 & 0 & 1 \\ 0 & 1 & 0 \end{pmatrix}.$$

The resultant matrix has eigenvalues, found in the usual way from the characteristic equation,

$$D, \tfrac{1}{2}\{D \pm \sqrt{D^2 + 4\alpha^2}\}.$$

This gives the form of energy level diagram shown in Figure 4 together with the corresponding eigenvectors. The magnetic behaviour in this case is rather different from applying a field parallel to the $z$-axis. Applying the field along the $x$-axis polarizes the ground state and induces a moment on it.

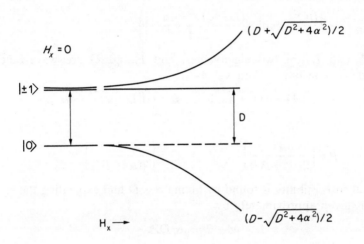

Figure 4. Energy levels of a $J = 1$ system with uniaxial crystal field showing how the levels split with a magnetic field applied along the $x$-axis

Since we have chosen the crystallographic $c$-axis as the axis of quantization, and the particular form for the crystal field energy given in equation (13), the required magnetic moment operator is $J_x$. Hence the magnetization is

$$M = -g\mu_B\langle\langle J_x\rangle\rangle. \tag{14}$$

From the eigenvectors and eigenvalues given in Figure 4 this expression is, for small non-zero fields $(0 < \alpha \ll D)$,

$$M = -\frac{g\mu_B}{Z}\left\{0 \cdot \exp\left[-\frac{D}{kT}\right] - \frac{4AP^2}{\alpha} \cdot \exp\left[-\frac{(D+\sqrt{D^2+4\alpha^2})}{2kT}\right]\right.$$

$$\left. -\frac{4BQ^2}{\alpha} \cdot \exp\left[-\frac{(D-\sqrt{D^2+4\alpha^2})}{2kT}\right]\right\}, \tag{15}$$

where $A$ and $B$ are two eigenvalues, and $P$ and $Q$ are wave-function normalization constants. Their values are

$$A = (\tfrac{1}{2})\{D - \sqrt{D^2+4\alpha^2}\}, \qquad B = (\tfrac{1}{2})\{D + \sqrt{D^2+4\alpha^2}\},$$

and

$$P = \left[\frac{\alpha^2}{2(\alpha^2+A^2)}\right]^{\frac{1}{2}}, \qquad Q = \left[\frac{2}{2(\alpha^2+B^2)}\right]^{\frac{1}{2}}.$$

The initial susceptibility is found by taking $\alpha \ll D$ and expanding the square root. It is given approximately by

$$M \sim 2g\mu_B\alpha/DZ, \tag{16}$$

where

$$Z = \exp[-D/kT] + \exp[-(D+\sqrt{D^2+4\alpha^2})/2kT]$$
$$+ \exp[-(D-\sqrt{D^2+4\alpha^2})/2kT].$$

The magnetization, although small, for small $\alpha$, is in fact much larger than the case when the field is applied along the $z$-axis.

## 2.2 Application to the real system

The behaviour of the simple $J = 1$ system gives the clue to the observed magnetization of Pr at low temperatures and high fields. The experimental magnetization, for fields along the [001] axis and the [110] axis is shown in Figure 5 and is taken from McEwan et al.[8] The $z$-axis of the model is the [001] axis, whereas the $x$-axis of Figure 1 corresponds to the [110] axis of the experimental work. For Pr the Landé $g$-factor is $(4/5)$, so therefore the saturated magnetic moment is 3.2 Bohr magnetons per atom. Applying the field along the $x$-axis causes the magnetization to rise rapidly, although,

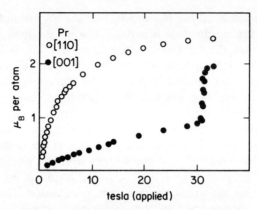

Figure 5. Experimentally observed magnetization of Pr single crystals with applied
magnetic field. From McEwen, Cock, Roeland, and Mackintosh (1973)

even in 30 T, the saturation value is not achieved. However, when the field
is applied along the $z$-axis there is only a small increase in magnetization
until an applied field of $\sim$32 T is reached, whereupon a large jump occurs in
the magnetization. This is due to a level crossing the $|0\rangle$ ground state. Unlike
our simple model, it is in fact the $|-3\rangle$ level crossing the $|0\rangle$ level that gives
this jump.

The computer program that is given at the end of this chapter calculates
the magnetization on each of the two crystallographic sites (i.e. the cubic
and hexagonal sites of the dhcp structure) for a given applied field and
temperature. This is achieved by setting up the matrix Hamiltonian of
equation (2) for the given applied field and diagonalizing it. The eigenvalues
and eigenvectors are then used in equations (14) or (8) to calculate a
magnetization at each site. This magnetization is converted to an internal
field by equation (1) and this new total internal field is compared with that
obtained on the previous iteration for the same applied field. If they are not
close enough the program iterates these steps until the self-consistent
solution is found. This is stored and a new field is chosen by adding on a
certain amount, decided from the input data, and the whole process repeated for a particular temperature. When all the field steps have been
calculated a table of results is printed on the line printer, then simple graphs
are drawn on the line printer of the magnetization of the cubic sites, the
hexagonal sites, and their average against the applied field. More temperatures may be requested and the entire process is repeated until a negative
temperature is encountered, whereupon the program terminates.

The experimental magnetization can be reproduced reasonably well by
choosing the values of the molecular field parameters $\lambda = -1.20$, $\mu = 4.14$ K,

together with the values of $B_2$, $B_4$, and $B_6$ that are set within the program. If these values of the parameters are used it will be found that the jump discontinuity is 'smoothed out' over some range $\Delta H$ where $\Delta H \sim (kT/g\mu_B)$. This is due to a hidden sophistication in that the free energy curve against magnetization has two minima separated by an energy barrier. At the critical field there are two allowed magnetizations, the smaller and the larger, and only if the temperature is sufficiently large can the system sample both and go smoothly from one to the other. For small temperatures the second state is not 'sensed' until sufficient energy, in the form of the applied field, has been delivered to drive the system over the potential barrier. In certain cases the program may fail to converge at this point as the allowed number of iterations, set within the program, is exceeded. The program then lists the results obtained up to that point and proceeds with the next temperature.

## 3. THE PROGRAM DATA

The necessary data to be presented to the program take the following form where each FORMAT refers to a new line of data. An integer number of field steps (FORMAT I3) is used. If this number is negative the graphs produced on the line printer will be approximately 7 in wide. This is a useful feature if the graphs are to be returned to a teletype, or are to form part of a report. The next data are the lowest field in tesla and the increment (FORMAT 2F6.2), two molecular field parameters $\lambda$, $\mu$ in kelvin (FORMAT 2F7.2) and an integer (FORMAT I1) that must be either a 0 or 1. If it is desired to apply the field along the $z$-axis use zero and if the field is to be applied along the $x$-axis use unity. Next there is an integer (FORMAT I1) which, if zero, inhibits the printing of all the eigenvalues and eigenvectors; it is this form that should generally be used. If the eigenvectors are desired, inputting unity here will print out nine eigenvalues across the page, followed underneath by the eigenvectors. These latter will each be columns of nine numbers and are to be interpreted as the coefficients of successively smaller $|J_z\rangle$ values. Thus, for example, in zero applied field, on the cubic sites the lowest eigenvalue would have below it (see Figure 2):

 0.000
 0.609
 0.000  which is interpreted as
 0.000  $0.609\,|{+}3\rangle - 0.509\,|0\rangle - 0.609\,|{-}3\rangle$
$-0.509$
 0.000
 0.000
$-0.609$
 0.000

The next numbers in the data are real and represent the required temperatures in kelvin (FORMAT F7.2). There can be any (reasonable) number of these, but the last one must be negative in order to stop the execution of the program.

## 4. A NOTE ON THE EIGENVECTORS

The eigenvectors are normalized so that, if $V_i$ represents a nine-tuple eigenvector, $V_i^T$ is its transpose, and $V_i^T V_i = 1$ for all $i$. They are also orthogonal in the sense that $V_i^T V_j = \delta_{ij}$. These properties are readily demonstrated in simple cases by examination of the eigenvectors produced by the program (to an accuracy of three decimal places). In a very large applied magnetic field the energy level scheme on both sites will tend towards nine levels, equally spaced, the lowest being the $|-4\rangle$ state, then $|-3\rangle$, and so on up to the $|+4\rangle$ state. If one was interested in a neutron-scattering experiment then the intensity of neutrons scattered with momentum change along the $z$-axis is proportional to the Boltzmann population of the initial energy level $|i\rangle$ and to the square of matrix elements connecting initial and final states $|f\rangle$, that is

$$I(E_{if}) \propto \frac{\exp(-E_i/kT)}{Z} \cdot |\langle i| J_z |f\rangle|^2,$$

where $E_{if}$ is the energy lost or gained by the neutron beam. With the subroutines provided with the program it is fairly simple to produce intensity spectra for different level schemes and different temperatures.

## 5. RUNNING THE PROGRAM

The data required to run the program and the input formats required were discussed in section 3. A useful starting point will be to use the program in an attempt to reproduce some recent experimental data (McEwan et al.[8]). The experiments were done at the temperature of liquid helium, 4.2 K, and with applied magnetic fields up to about 34 T. These fields were applied to the $x$- and $z$-axes in turn. This range of applied magnetic field represents the limit of high field experimental equipment currently available. Fields in excess of this may be applied in the program. It may be interesting to determine at what field you would expect to obtain magnetic saturation of a pure crystal of praseodymium. Choosing molecular field parameters $\lambda = -1.20$, $\mu = 4.14$ K and a temperature of 4.2 K gives reasonable representation of the experimental results. For convenience, only the regions where the curves begin to become asymtotic are drawn at the end of the chapter, although the computer program will always give complete curves, starting at 0 tesla. The deviations from experiment are discussed briefly in section 2.2.

Keeping $\lambda$ and $\mu$ fixed, for the present, the temperature should be varied. By pumping on the vapour above the liquid helium, temperatures down to 1 K are possible experimentally so temperatures down to this should be tried in the simulation. Lower temperatures may also be used, the effect being to sharpen up the transition that occurs for magnetic fields applied parallel to the $z$-axis. However, even though a temperature as low as a few millikelvin ($\sim$50 mK) may be produced by using the technique of magnetic cooling (a good description may be found in Mendelssohn[9]), temperatures as low as this are not recommended for the program for two reasons. Firstly, at these very low temperatures some new physics occurs; the magnetic moments of the nuclei may become aligned through the hyperfine interaction and would, therefore, add an extra contribution to the molecular field. The approximations made in setting up the model used here are inadequate to represent this effect correctly. Secondly, overflows may occur when calculating the exponential factors, thus causing the program to end abnormally, the exact effect will depend on the computer operating system under which the job is being run.

Higher temperatures are also well worth investigating. The next convenient experimental temperature would be 77 K, the temperature of liquid nitrogen. Increasing the temperature allows more of the energy levels to be populated in proportion to the usual Boltzmann factor. This changes the averaged magnetic moment of each ion and smears out the magnetic ordering over a greater range of applied magnetic fields. For higher temperatures, stronger magnetic fields will be required to induce the ordering. Note that praseodymium melts at about 1300 K so temperatures larger than this would not be sensible.

Having investigated the role of temperature, the other parameters easily varied are the two molecular field constants $\lambda$ and $\mu$. Equation (1) of section 2 shows how these parameters are related to the strengths of the internal magnetic fields. By applying a large hydrostatic pressure to the crystal one might expect to change the lattice spacings and hence alter $\lambda$ and $\mu$. Because the interaction between two local magnetic moments has the oscillatory RKKY form, changing the lattice spacing by small amounts may change both the magnitude and the sign of either $\lambda$ or $\mu$, or both. Unfortunately, the way in which this would happen is rather complex, and lies on the frontiers of current research, so it is difficult to give an easy physical picture. Nevertheless it is instructive to vary both $\lambda$ and $\mu$. Try varying them one at a time, in small steps, to build up a picture of the behaviour of the system you have created within a parameter space limited by, say, $-5 \leqslant \lambda$, $\mu \leqslant 5$ K. This will represent a lot of computing and may well need more than one session to do it properly.

As an extension to the work, there is the possibility of building up the sort of pictures you would get from a neutron diffraction experiment. This was discussed briefly in section 4.

Not only are there anomalies in the magnetization curves, there are also associated anomalies in the resistivity curves or magneto-resistance, if it is measured as a function of applied magnetic field. For those who would like to study this aspect, and who are willing to write their own programs it is an easy matter. The equations necessary may be found in either of the two references[10] at the end of this chapter, the only extra ones required are those to calculate two matrices that contain the scattering probability information.

Another extension of the work, which requires no extra programming, is to keep $\lambda = -1.20$, $\mu = 4.14$ and to reset the constants BH2, BH4, BH6, BC2, BC4, BC6 used in the program. These are the constants multiplying the various Stevens operators and represent the strengths of each of the terms of different symmetry in the electric crystal field. Altering their values will have the effect of altering the energy levels and the order in which they occur in the absence of an applied magnetic field. Thus a hypothetical praseodymium may be 'produced' in which a completely different arrangement of the levels is investigated. This extension will be most suitable for those who wish to delve into the current research literature to obtain the most up-to-date values for the crystal field constants in a point charge model.

## REFERENCES

1. M. A. Rudermann and C. Kittel, *Phys. Rev.*, **96,** 99 (1954).
2. T. Kasuya, *Prog. Theor. Phys.*, **16,** 45 (1956).
3. K. Yosida, *Phys. Rev.*, **106,** 893 (1957).
4. P. Weiss, *J. de Phys.*, **6,** 661 (1907).
5. K. W. H. Stevens, *Proc. Phys. Soc. (Lond.)*, **A65,** 209 (1952).
6. M. T. Hutchings, *Solid State Phys.*, **16,** 227 (1964).
7. R. H. Dicke, and J. P. Wittke, *Introduction to Quantum Mechanics* (Addison-Wesley, Reading, Mass., 1961).
8. K. A. McEwan, G. J. Cock, L. W. Roeland, and A. R. Mackintosh, *Phys. Rev. Lett.*, **30,** 287 (1973).
9. K. Mendelssohn, *The Quest for Absolute Zero* (Weidenfeld and Nicolson, London, 1966).
10. J. A. G. Temple, K. A. McEwan, 'Magneto-resistance of praseodymium' in *Crystal Field Effects in Metals*, Ed. A. Furrer (Plenum, New York, 1977); K. A. McEwan, J. A. G. Temple, and G. D. Webber, *Physica*, **86–88**(B), 533 (1977).

PRASEODYMIUM

```
      DIMENSION RHH(9,9),RHC(9,9),EVALC(9),EVALH(9)
      DIMENSION RZO(9,9),RMO(9,9),RPO(9,9),RES(50,5)
      DIMENSION RXO(9,9),USE(9)
      LOGICAL A4
      A4=.FALSE.
CCCCCCCCCCCCCCCCCCCCCCCCCCCCCCCCCCCCCCCCCCCCCCCCCCCCCCCCCCCCCCCCCCCCCC
C
C    THE LOGICAL VARIABLE A4 GOVERNS THE SIZE OF THE GRAPHS PRODUCED ON
C 120 COLUMNS THAT IS 12 INCHES. A4=TRUE GIVES GRAPHS OCCUPYING
C  ONLY 71 COLUMNS WIDTH AND IS THEREFORE SUITABLE FOR OUTPUTTING
C RESULTS TO NORMAL TELETYPES
C  A4 IS SET TRUE IF THE NUMBER OF FIELD STEPS (THE FIRST DATA ITEM)
C  IS NEGATIVE, OTHERWISE IT REMAINS FALSE.
C
CCCCCCCCCCCCCCCCCCCCCCCCCCCCCCCCCCCCCCCCCCCCCCCCCCCCCCCCCCCCCCCCCCCCCC
C
C SET THE OPERATOR MATRICES TO ZERO BEFORE SETTING THEIR
C  NON ZERO ELEMENTS
C
      DO 100 I=1,9
      DO 100 J=1,9
      RZO(I,J)=0.0
      RMO(I,J)=0.0
      RXO(I,J)=0.0
      RPO(I,J)=0.0
  100 CONTINUE
C
C SET THE NON ZERO ELEMENTS OF THE FOUR OPERATOR MATPICES
C
      DO 200 I=1,9
      FI=FLOAT(5-I)
      RZO(I,I)=FI
  200 CONTINUE
      DO 205 I=2,9
      J=I-1
      FJ=FLOAT(J)
      FI=FLOAT(10-I)
      XX=FI*FJ
      XX=SQRI(XX)
      RPO(J,I)=XX
      RMO(I,J)=XX
  205 CONTINUE
      DO 210 I=1,9
      DO 210 J=1,9
      RXO(I,J)=0.50*(RPO(I,J)+RMO(I,J))
  210 CONTINUE
CCCCCCCCCCCCCCCCCCCCCCCCCCCCCCCCCCCCCCCCCCCCCCCCCCCCCCCCCCCCCCCCCCCCCC
C
C DATA IS READ IN WITH THE FOLLOWING FORMATS
C
C NUMBER OF FIELD STEPS FOR EACH TEMPERATURE   FORMAT   I3
C  IF NUMBER OF FIELD STEPS IS NEGATIVE
C  THEN THE LINEPRINTER GRAPHS ARE PRODUCED A4 SIZE.
C
C LOWEST FIELD AND FIELD INCREMENT IN TESLA    FORMAT 2F6.2
C  TWO MOLECULAR FIELD PARAMETERS IN KELVIN    FORMAT 2F7.2
```

PRASEODYMIUM

```
C (FOR DETAILS OF MOLECULAR FIELD CONSTANTS SEE EQN.(1) AND ASSOCIATED
C     TEXT).
C
C INTEGER=0  FOR FIELD PARALLEL TO THE Z-AXIS FORMAT  I1
C       =1  FOR FIELD PARALLEL TO THE X-AXIS
C INTEGER=0  FOR NO PRINTING OF EIGENVECTORS FORMAT    I1
C       =1  FOR PRINTING OF ALL EIGENVECTORS
C SERIES OF TEMPERATURES THE LAST OF WHICH MUST
C  BE NEGATIVE.                              FORMAT  F7.2
C
C
CCCCCCCCCCCCCCCCCCCCCCCCCCCCCCCCCCCCCCCCCCCCCCCCCCCCCCCCCCCCCCCCCCCCCCCCCC
C
      READ(1,1008) NHS
      IF(NHS.LT.0) A4=.TRUE.
      NHS=IABS(NHS)
      IF(NHS.GT.50) NHS=50
      READ(1,1010) PHL,RHS
      READ(1,1011) RM1,RM2
      READ(1,1009) ID
      READ(1,1009) IP
      WRITE(2,1200)
      WRITE(2,1201) RHL
      WRITE(2,1202) NHS,RHS
      IF(ID.EQ.0) GOTO 220
      WRITE(2,1203)
      GOTO 230
  220 WRITE(2,1204)
  230 CONTINUE
      WRITE(2,1205) RM1,RM2
CRYSTAL FIELD PAPAMETER ARE SET
      BH2=4.920
      BH4=1.390
      BH6=1.240
      BC2=0.0
      BC4=BH4
      BC6=BH6
  300 READ(1,1012) T
      IF(T.LT.0.0) GOTO 940
      IM=30
      DO 120 I=1,5
      DO 120 J=1,50
      RES(J,I)=0.0
  120 CONTINUE
      IH=0
      RMAGC=0.0
      RMAGH=0.0
CU IS CONVERSION FACTOR FROM TESLA TO KELVIN
      CU=(0.80*0.92730)/1.38060
      IF(RHL.GT.0.0) IM=100
  310 IT=0
      IH=IH+1
  320 IT=IT+1
      IF(IT.LT.IM GOIO 325
      WRITE(2,2900)
      WRITE(2,2901) IH,II,IM
```

PRASEODYMIUM


```
      GOTO 900
  325 IF(IH.GT.NHS) GOTO 900
      IF(IH.GT.S0) GOTO 900
      FIH=FLOAT(IH-1)
CALCULATE APPLIED MAGNETIC FIELD RHA
      RHA=RHL+FIH*RHS
      IF(ID.EQ.0) GOTO 330
C
CONSTPUCT THE HAMILTONIAN MATRICES FOR THE ENERGY OF THE IONS
CONSIDERING THE INTERNAL MAGNETIC FIELD AND THE
CRYSTAL FIELD ENERGY LEVELS. UNITS ARE KELVIN.
C
      HZC=0.0
      HZH=0.0
      HXC=(RHA+PM1*RMAGC+RM2*RMAGH)*CU*0.50
      HXH=(RHA+RM1*RMAGH+RM2*RMAGC)*CU*0.50
      GOTO 340
  330 HXC=0.0
      HXH=0.0
      HZC=(RHA+RM1*RMAGC+RM2*RMAGH)*CU
      HZH=(RHA+RM1*RMAGH+RM2*RMAGC)*CU
  340 CALL PRH(RHH,0,HXH,HZH,BH2,BH4,BH6)
      CALL PRH(RHC,1,HXC,HZC,BC2,BC4,BC6)
CCCCCCCCCCCCCCCCCCCCCCCCCCCCCCCCCCCCCCCCCCCCCCCCCCCCCCCCCCCCCCCCCCCCCCC
C
CALCULATE THE EIGENVALUES AND EIGENVECTORS OF THE TWO MATRICES
C
C     ROUTINES F01AJF+F02AMF ARE STANDARD LIBRARY ROUTINES(NAG)
C     WHICH CALCULATE EIGENVALUES AND EIGENVECTORS OF REAL
C     SYMMETRIC MATRICES BY HOUSEHOLDER REDUCTION TO TRI-
C     DIAGONAL FORM FOLLOWED BY THE QR ALGORITHM TO COMPLETE
C     THE DIAGONALISATION.
C     FOR DETAILS OF HOUSEHOLDER AND QP ALGORITHMS SEE BOOKS ON
C     NUMERICAL METHODS, FOR EXAMPLE ACTON R.S. "NUMERICAL METHODS
C     THAT WORK" PUBLISHED BY HARPER,N.Y. (1970) PAGE 347
C
C
C THE PARAMETER LIST OF F01AJF(N,TOL,A,IA,D,E,Z,IZ) IS
C     N=INTEGER,THE OPDER OF MATRIX A
C     TOL=REAL,MACHINE DEPENDENT CONSTANT, FOR ICL 1900 TOL=2.0**(-218)
C     A=REAL ARRAY,OF DIMENSION AT LEAST(N,N) CONTAINING THE SYMMETRIC
C            MATRIX,THE LOWER TRIANGLE ONLY IS REQUIRED. THE ARRAY IS
C            NOT OVERWRITTEN BY THE ROUTINE.
C     IA=INTEGER,THE FIRST DIMENSION OF A AS DEFINED IN THE CALLING.
C     D=REAL ARRAY,OF DIMENSION AT LEAST (N), ON EXIT IT CONTAINS THE
C            DIAGONAL ELEMENTS OF TRIDIAGONAL MATRIX.
C     E=REAL ARRAY,OF DIMENSION AT LEAST (N), ON EXIT IT CONTAINS THE
C            N-1 OFF DIAGONAL ELEMENTS OF TPIDIAGONAL MATRIX.
C     Z=REAL ARRAY,OF DIMENSION AT LEAST (N,N). ON EXIT IT CONTAINS
C            THE ORTHOGONAL MATRIX Q THE PRODUCT OF THE HOUSEHOLDEP
C            TRANSFORMATIONS.
C     IZ=INTEGER,THE FIRST DIMENSION OF Z AS DEFINED IN CALLING SEG.
C
C
C THE PARAMETER LIST OF F02AMF(N,ACC,D,E,Z,IZ,IFAIL) IS
C     N=INTEGER,THE ORDER OF TRIDIAGONAL MATRIX T.
```

PRASEODYMIUM

```
C      ACC=REAL,SMALLEST NUMBER ON THE COMPUTER SUCH THAT 1+ACC=1
C           (ON ICL 1900 ACC=2.0**(-37))
C      D=REAL ARRAY,OF DIMENSION AT LEAST (N),CONTAINING DIAGONAL
C           ELEMENTS OF T.
C      E=REAL ARRAY,OF DIMENSION AT LEAST (N) CONTAINING THE SUB-
C           DIAGONAL ELEMENTS OF T STORED IN E(2)-=E(N). IT IS
C           OVERWRITTEN BY THE ROUTINE.
C      Z=REAL ARRAY, DIMENSION AT LEAST (N,N), IF EIGENVECTORS OF THE
C           FULL SYMMETRIC MATRIX ARE REQUIRED IT SHOULD CONTAIN THE
C           Q (SEE F01AJF). ON EXIT IT CONTAINS THE NORMALIZED
C           EIGENVECTORS SUCH THAT Z(I,J),I=1,N CORRESPONDS TO
C           EIGENVALUE J.
C      IZ=INTEGER,THE FIRST DIMENSION OF Z AS DECLARED.
C      IFAIL=INTEGER FLAG, GOVERNS ON ENTRY THE TYPES OF FAILURE
C           THAT WILL BE DETECTED,ON EXIT THE TYPE OF FAILURE.
C           IFAIL=0 ON EXIT FOR SUCCESSFUL COMPLETION.
C
C****************************************************************************
C
C NOTE THAT IF NAG LIBRARY ROUTINES ARE NOT AVAILABLE THEN THESE
C     ROUTINES HAVE THEIR EXACT EQUIVALENTS IN MOST SCIENTIFIC
C     LIBRARIES( UNDER DIFFERENT NAMES AND WITH, POSSIBLY, DIFFERENT
C     ARGUMENT LISTS)
C****************************************************************************
C THOSE INTERESTED IN NUMERICAL EIGENVALUE PROBLEMS SHOULD CONSULT
C MARTIN R.S.,REINSCH C.,WILKINSON J.H.,NUM.MATH.BAND(11)181-95(1968)
C  BOWDLER H,MARTIN R.S.,REINSCH C.,WILKINSON J.H,
C                          NUM MATH BAND(11) 293-306 (1968)
C****************************************************************************
       IFAIL=0
       CALL F01AJF(9,2.0**(-218),RHH,9,EVALH,USE,PHH,9)
       CALL F02AMF(9,2.0**(-37),EVALH,USE,PHH,9,IFAIL)
       IF(IFAIL.EQ.0) GOTO 345
       WRITE(2,2000)
       IFAIL=0
  345 CONTINUE
       CALL F01AJF(9,2.0**(-218),RHC,9,EVALC,USE,RHC,9)
       CALL F02AMF(9,2.0**(-37),EVALC,USE,RHC,9,IFAIL)
       IF(IFAIL.EQ.0) GOTO 346
       WRITE(2,2000)
       IFAIL=0
  346 CONTINUE
       IF(IP.EQ.0) GOTO 349
       IF(IT.GT.1) GOTO 349
       WRITE(2,2005)
       IF(ID.EQ.0) GOTO 347
       WRITE(2,2006) RHA
       GOTO 348
  347 WRITE(2,2007)RHA
  348 CONTINUE
       WRITE(2,2001)
C
C  PRINT EIGENVALUES AND EIGENVECTORS IF REQUIRED
C
       IE=1
       IO=9
```

PRASEODYMIUM

```
      CALL PRINTM(EVALH,1,9,IE)
      WRITE(2,2002)
      CALL PRINTM(RHH,9,9,IU)
      WRITE(2,2003)
      CALL PRINTM(EVALC,1,9,IE)
      WRITE(2,2002)
      CALL PRINTM(RHC,9,9,IU)
  349 CONTINUE
C
CALCULATE THE MAGNETIZATION ON BOTH SITES
C
      IF(ID.EQ.0) GOTO 350
      CALL PRT(9,RHC,EVALC,T,RXO,USE1)
      CALL PRI(9,RHH,EVALH,T,RXO,USE2)
      GOTO 360
  350 CALL PRT(9,RHC,EVALC,T,RZO,USE1)
      CALL PRT(9,RHH,EVALH,T,RZO,USE2)
  360 USE1=-0.80*USE1
      USE2=-0.80*USE2
C
CHECK FOR CONVERGENCE OF MAGNETIZATION
C
      XX=ABS(USE1-RMAGC)
      YY=ABS(USE2-RMAGH)
      IF(XX.GT.0.20) GOTO 390
      IF(YY.LE.0.20) GOTO 400
  390 RMAGC=USE1
      RMAGH=USE2
      GOTO 320
  400 CONTINUE
C
C STORE RESULTS IN ARRAY RES
C
      RES(IH,1)=T
      RES(IH,2)=RHA
      RES(IH,3)=USE1
      RES(IH,4)=USE2
      RES(IH,5)=0.50*(USE1+USE2)
C
CHECK WHETHER REQUISITE NUMBER OF FIELD STEPS HAVE BEEN
COMPLETED AND IF SO OUTPUT THE RESULTS TO LINEPRINTER
C
      IF(IH.LT.NHS) GOTO 310
  900 CONTINUE
C
CONSTRUCT A TABLE OF RESULTS
C
      WRITE(2,2004)
      WRITE(2,2008)
      CALL PRINTM(RES,50,5,NHS)
C
CONSTRUCT A GRAPHICAL DISPLAY ON THE LINEPRINTER
C
      CALL PRG(RES,50,5,IH,2,3,T,A4)
      CALL PRG(RES,50,5,IH,2,4,T,A4)
      CALL PRG(PES,50,5,IH,2,5,T,A4)
```

PRASEODYMIUM

```
        GOTO 300
  940 STOP
 1009 FORMAT(I1)
 1010 FORMAT(2F6.2)
 1012 FORMAT(F7.2)
 1008 FORMAT(I3)
 1011 FORMAT(2F7.2)
 2008 FORMAT(1H0,3X,6HTEMP K,5X,5HFIELD,5X,5HMAG C,5X,5HMAG H,5X
     A,7HAVERAGE)
 2000 FORMAT(1H0,22HFAILURE IN NAG F02 AMF)
 2001 FORMAT(1H0,37HEIGENVALUES HEXAGONAL SITES IN KELVIN)
 2002 FORMAT(1H0,26HCORRESPONDING EIGENVECTORS)
 2003 FORMAT(1H0,33HEIGENVALUES CUBIC SITES IN KELVIN)
 2004 FORMAT(1H1,11HRESULTS ARE)
 2005 FORMAT(1H0,25HAPPLIED FIELD IN TESLA IS)
 2006 FORMAT(1H+,28X,F8.2,14HIN X DIRECTION)
 2007 FORMAT(1H+,28X,F8.2,14HIN Z DIRECTION)
 2900 FORMAT(1H0,31HMAGNETIZATION IS NOT CONVERGING)
 2901 FORMAT(1H0,3HIH=,I3,3HIT=,I3,3HIM=,I3)
 1200 FORMAT(1H1,27HINPUT DATA FOR PRASEODYMIUM)
 1201 FORMAT(1H0,15HLOWEST FIELD IS,F8.2,6H TESLA)
 1202 FORMAT(1H0,9HTHERE ARE,I3,9H STEPS OF,F8.2,6H TESLA)
 1203 FORMAT(1H+,40X,22HAPPLIED IN X DIRECTION)
 1204 FORMAT(1H+,40X,22HAPPLIED IN Z DIRECTION)
 1205 FORMAT(1H0,32HMOLECULAR FIELD PARAMETERS USED=,2(F7.2,3X))
      END
      SUBROUTINE PRG(ROUT,NN,MM,NP,IX,IY,TT,A4)
      DIMENSION ROUT(NN,MM),ILINE(120),ICHAR(3)
      LOGICAL A4
      DATA ICHAR(1),ICHAR(2),ICHAR(3)/1H ,1H.,1H*/
      IF(NP.GT.NN) GOTO 900
CCCCCCCCCCCCCCCCCCCCCCCCCCCCCCCCCCCCCCCCCCCCCCCCCCCCCCCCCCCCCCCCCCCCCC
C
CODE OUTPUTS A SIMPLE GRAPH OF ROUT ONTO LINEPRINTER
C
CCCCCCCCCCCCCCCCCCCCCCCCCCCCCCCCCCCCCCCCCCCCCCCCCCCCCCCCCCCCCCCCCCCCCC
      ILIM=120
      IF(A4) ILIM=71
      SMALL=10.0**(-10)
      RYM=ROUT(1,IY)
      RYS=0.0
      DO 100 II=1,NP
      IF(ROUT(II,IY).GT.RYM) RYM=ROUT(II,IY)
      IF(ROUT(II,IY).LT.RYS) RYS=ROUT(II,IY)
  100 CONTINUE
      YB=RYS
      AYB=ABS(YB)
      IF(AYB.LT.SMALL) YB=0.0
      AYB=ABS(RYM-YB)
      IF(AYB.LT.SMALL) GOTO 900
      IF(.NOT.A4) RYS=109.0/(RYM-YB)
      IF(A4) RYS=69.0/(RYM-YB)
      IF(IY.EQ.3) GOTO 110
      IF(IY.EQ.4) GOTO 112
      WRITE(2,1902)
      GOTO 116
```

PRASEODYMIUM

```
  110 WRITE(2,1900)
      GOTO 116
  112 WRITE(2,1901)
  116 CONTINUE
      WRITE(2,1903) TT
      IF(IX.EQ.2) GOTO 120
      IF(IX.EQ.1) GOTO 122
      GOTO 130
  120 WRITE(2,2000)
      GOTO 130
  122 WRITE(2,2002)
  130 CONTINUE
  140 WRITE(2,2003)
      DO 320 II=1,ILIM
  320 ILINE(II)=ICHAR(2)
      IF(.NOT.A4) WRITE(2,2500) YB,RYM
      IF(A4) WRITE(2,2503) YB,RYM
      WRITE(2,2501)(ILINE(II),II=1,ILIM)
      RX=55.0/FLOAT(NP)
      ILX=IFIX(RX-0.50)
      DO 400 II=1,NP
      DO 360 JJ=1,ILIM
      ILINE(JJ)=ICHAR(1)
  360 CONTINUE
      ILINE(10)=ICHAR(2)
      Y=(ROUT(II,IY)-YB)*RYS
      Y=Y+0.50
      ILY=IFI/(Y)+10
      IF(ILY.GT.ILIM) ILY=ILIM
      ILINE(ILY)=ICHAR(3)
      WRITE(2,2501)(ILINE(JJ),JJ=1,ILIM)
      WRITE(2,2502) ROUT(II,IX)
      ILINE(ILY)=ICHAR(1)
      ILINE(10)=ICHAR(2)
      IF(ILX.LE.1) GOTO 390
      DO 380 JJ=1,ILX
      WRITE(2,2501)(ILINE(KK),KK=1,ILIM)
  380 CONTINUE
  390 CONTINUE
  400 CONTINUE
 1900 FORMAT(1H1,20X,11HCUBIC SITES)
 1901 FORMAT(1H1,20X,15HHEXAGONAL SITES)
 1902 FORMAT(1H1,20X,23HAVERAGE FROM BOTH SITES)
 1903 FORMAT(1H ,40X,14HTEMPERATURE IS,F8.2,7H KELVIN)
 2000 FORMAT(1H0,12H FIELD TESLA)
 2002 FORMAT(1H0,11H TEM KELVIN)
 2003 FORMAT(1H+,50X,17HMAGNETISATION(BM))
C BM=BOHR MAGNETONS
 2500 FORMAT(1H ,10X,E10.2,90X,E10.2)
 2501 FORMAT(1H ,120A1)
 2502 FORMAT(1H+,F9.3)
 2503 FORMAT(1H ,10X,E10.2,28X,E10.2)

  900 RETURN
      END
      SUBROUTINE PRH(RH,CH,HX,HZ,B2,B4,B6)
```

PRASEODYMIUM

```
      DIMENSION RH(9,9)
      REAL B2,B4,B6,HX,HZ,H,G,X,XS,MU,SQP
      INTEGER CH
C
CCCCCCCCCCCCCCCCCCCCCCCCCCCCCCCCCCCCCCCCCCCCCCCCCCCCCCCCCCCCCCCCCCCCCC
C
CONSTRUCTS THE HAMILTONIAN MATRICES FOR PRASEODYMIUM AND STORES IN RH
CH IS=0 FOR HEXAGONAL SITES AND=1 FOR CUBIC SITES
CONSTANTS B2 B4 B6 ARE THE THREE CRYSTAL FIELD PARAMETERS
CODE USES HX AND HZ FOR THE TOTAL MAGNETIC FIELDS IN X AND Z DIRE
CTIONS RESPECTIVELY.
C
      DO 100 I=1,9
      DO 100 J=1,9
  100 RH(I,J)=0.0
      RH(1,1)=28.0*B2+14.0*B4+4.0*B6+4.0*HZ
      RH(1,2)=SQRT(8.0)*HX
      RH(2,1)=RH(1,2)
      RH(8,9)=RH(1,2)
      RH(9,8)=RH(8,9)
      RH(1,7)=SQRT(7.0)*5.50*B6
      RH(7,1)=RH(1,7)
      RH(3,9)=RH(1,7)
      RH(9,3)=RH(3,9)
      RH(2,2)=7.0*B2-21.0*B4-17.0*B6+3.0*HZ
      RH(3,2)=SQRT(14.0)*HX
      RH(2,3)=RH(3,2)
      RH(7,8)=RH(2,3)
      RH(8,7)=RH(7,8)
      RH(2,8)=(77.0*B6)/4.0
      RH(8,2)=RH(2,8)
      RH(3,3)=-8.0*B2-11.0*B4+22.00*B6+2.0*HZ
      RH(3,4)=SQRT(18.0)*HX
      RH(4,3)=RH(3,4)
      RH(6,7)=RH(3,4)
      RH(7,6)=RH(6,7)
      RH(4,4)=-17.0*B2+9.0*B4+B6+HZ
      RH(4,5)=SQRT(20.0)*HX
      RH(5,4)=RH(4,5)
      RH(5,6)=RH(4,5)
      RH(6,5)=RH(5,6)
      RH(5,5)=-20.0*B2+18.0*B4-20.0*B6
      RH(6,6)=RH(4,4)-2.0*HZ
      RH(7,7)=RH(3,3)-4.0*HZ
      RH(8,8)=RH(2,2)-6.0*HZ
      RH(9,9)=RH(1,1)-8.0*HZ
      IF(CH.EQ.0) GOTO 900
C
CUBIC SYMMETRY TERMS ADDED NOW
C
      RH(1,4)=SQRT(7.0)*(10.0*B4-5.0*B6)
      RH(4,1)=RH(1,4)
      RH(2,5)=SQRT(70.0)*(3.0*B4+1.250*B6)
      RH(5,2)=RH(2,5)
      RH(3,6)=10.0*B4+8.750*B6
      RH(6,3)=RH(3,6)
```

PRASEODYMIUM

```
      RH( 6,9 )=-RH( 1,4 )
      RH( 9,6 )=RH( 6,9 )
      RH( 5,8 )=-RH( 2,5 )
      RH( 8,5 )=RH( 5,8 )
      RH( 4,7 )=-RH( 3,6 )
      RH( 7,4 )=RH( 4,7 )
  900 RETURN
      END
      SUBROUTINE PRINTM( RMAT,I1,I2,IPOW )
      DIMENSION PMAI( I1,I2 )
C
COLUMNS OF MATRIX RMAT( I1,I2 ) PRINTED BY SUBROUTINE ( MAX 12 COLS )
C
      LIM=12
      IF( I2.GT.LIM ) LIM=12
      IF( IROW.GT.I1 ) GOTO 910
      WRITE( 2,2001 )
      DO 100 II=1,IROW
      WRITE( 2,2000 )( RMAT( II,JJ ),JJ=1,LIM )
  100 CONTINUE
      GOTO 900
  910 WRITE( 2,2003 ) IROW,I1
 2000 FORMAT( 1H ,12F10.4 )
 2001 FORMAT( 1H0 )
 2003 FORMAT( 1H0,24HPRINTM CALLED WITH IROW=,I6,5H  I1=,I6 )
  900 RETURN
      END
      SUBROUTINE PRT( NN,EVEC,EVAL,TEM,ROP,RES )
      DIMENSION EVEC( NN,NN ),EVAL( NN ),ROP( NN,NN )
      DIMENSION RMOM( 20 ),RU( 20 )
      REAL NUM,DEN
      INTEGER RR
      IF( NN.GT.20 ) GOTO 900
CCCCCCCCCCCCCCCCCCCCCCCCCCCCCCCCCCCCCCCCCCCCCCCCCCCCCCCCCCCCCCCCCCCCC
C
CALCULATES THE THERMAL AVERAGE OF AN OPERATOR ROP  GIVEN THE
CORECT EIGENVALUES STORED IN ASCENDING ORDER IN EVAL AND THE
CORRESPONDING EIGENVECTORS STORED IN EVEC.  TEM IS THE TEMPERATURE.
C
CCCCCCCCCCCCCCCCCCCCCCCCCCCCCCCCCCCCCCCCCCCCCCCCCCCCCCCCCCCCCCCCCCCCC
      NUM=0.0
      DEN=0.0
      IF( TEM.LT.0.000000001 ) GOTO 800
      DO 100 II=1,20
      RMOM( II )=0.0
      RU( II )=0.0
  100 CONTINUE
      DO 200 KK=1,NN
      DO 110 II=1,NN
  110 RU( II )=0.0
      DO 160 II=1,NN
      DO 140 JJ=1,NN
      RU( II )=RU( II )+ROP( II,JJ )*EVEC( JJ,KK )
  140 CONTINUE
  160 CONTINUE
      DO 170 II=1,NN
```

PRASEODYMIUM

```
      RMOM(KK)=RMOM(KK)+EVEC(II,KK)*RU(II)
170 CONTINUE
200 CONTINUE
      NUM=EVAL(1)/TEM
      IF(NUM.LE.-60.0) GOTO 400
      NUM=0.0
      DO 210 II=1,NN
      XX=-EVAL(II)/TEM
      DEN=DEN+EXP(XX)
      NUM=NUM+RMOM(II)*EXP(XX)
210 CONTINUE
      RES=NUM/DEN
      GOTO 900
400 RR=0
      NUM=0.0
      DO 420 II=1,NN
420 IF(EVAL(II).EQ.EVAL(1)) RR=RR+1
      DO 430 II=1,RR
430 NUM=NUM+RMOM(II)
      JJ=RR+1
      XX=EVAL(RR)-EVAL(JJ)
      XX=EXP(XX/TEM)
      NUM=NUM+XX
      DEN=1.0+XX
      PES=NUM/DEN
      GOTO 900
800 DO 820 II=1,20
      RMOM(II)=0.0
820 RU(II)=0.0
      DO 830 II=1,NN
      DO 825 JJ=1,NN
825 RU(II)=RU(II)+ROP(II,JJ)*EVEC(JJ,1)
830 CONTINUE
      DO 840 II=1,NN
      RMOM(1)=RMOM(1)+EVEC(II,1)*RU(II)
840 CONTINUE
      RES=RMOM(1)

      END
```

X DIRECTION RESULTS

INPUT DATA FOR PRASEODYMIUM
LOWEST FIELD IS     0.00 TESLA
THERE ARE 40 STEPS OF     2.00 TESLA

                                      APPLIED IN X DIRECTION
MOLECULAR FIELD PARAMETERS USED=  -1.20     4.14
RESULTS ARE

| TEMP K | FIELD | MAG C | MAG H | AVERAGE |
|---|---|---|---|---|
| 4.2000 | 0.0000 | 0.0000 | -0.0000 | 0.0000 |
| 4.2000 | 2.0000 | 0.2789 | 0.5438 | 0.4113 |
| 4.2000 | 4.0000 | 0.5092 | 1.2042 | 0.8567 |
| 4.2000 | 6.0000 | 0.7421 | 1.6494 | 1.1957 |
| 4.2000 | 8.0000 | 0.9151 | 1.9961 | 1.4556 |
| 4.2000 | 10.0000 | 1.0634 | 2.2234 | 1.6434 |
| 4.2000 | 12.0000 | 1.1603 | 2.3644 | 1.7624 |
| 4.2000 | 14.0000 | 1.2907 | 2.5025 | 1.8966 |
| 4.2000 | 16.0000 | 1.3748 | 2.5829 | 1.9788 |
| 4.2000 | 18.0000 | 1.4722 | 2.6660 | 2.0691 |
| 4.2000 | 20.0000 | 1.5455 | 2.7167 | 2.1311 |
| 4.2000 | 22.0000 | 1.6143 | 2.7593 | 2.1868 |
| 4.2000 | 24.0000 | 1.6897 | 2.8080 | 2.2488 |
| 4.2000 | 26.0000 | 1.7500 | 2.8376 | 2.2938 |
| 4.2000 | 28.0000 | 1.8067 | 2.8635 | 2.3351 |
| 4.2000 | 30.0000 | 1.8601 | 2.8862 | 2.3732 |
| 4.2000 | 32.0000 | 1.9149 | 2.9141 | 2.4145 |
| 4.2000 | 34.0000 | 1.9622 | 2.9313 | 2.4467 |
| 4.2000 | 36.0000 | 2.0069 | 2.9467 | 2.4768 |
| 4.2000 | 38.0000 | 2.0493 | 2.9607 | 2.5050 |
| 4.2000 | 40.0000 | 2.0895 | 2.9734 | 2.5315 |
| 4.2000 | 42.0000 | 2.1289 | 2.9894 | 2.5591 |
| 4.2000 | 44.0000 | 2.1651 | 2.9996 | 2.5824 |
| 4.2000 | 46.0000 | 2.1997 | 3.0090 | 2.6044 |
| 4.2000 | 48.0000 | 2.2327 | 3.0177 | 2.6252 |
| 4.2000 | 50.0000 | 2.2642 | 3.0257 | 2.6449 |
| 4.2000 | 52.0000 | 2.2943 | 3.0332 | 2.6637 |
| 4.2000 | 54.0000 | 2.3232 | 3.0401 | 2.6816 |
| 4.2000 | 56.0000 | 2.3507 | 3.0492 | 2.6999 |
| 4.2000 | 58.0000 | 2.3773 | 3.0550 | 2.7162 |
| 4.2000 | 60.0000 | 2.4028 | 3.0605 | 2.7317 |
| 4.2000 | 62.0000 | 2.4275 | 3.0657 | 2.7466 |
| 4.2000 | 64.0000 | 2.4511 | 3.0705 | 2.7608 |
| 4.2000 | 66.0000 | 2.4740 | 3.0751 | 2.7745 |
| 4.2000 | 68.0000 | 2.4960 | 3.0794 | 2.7877 |
| 4.2000 | 70.0000 | 2.5173 | 3.0834 | 2.8004 |
| 4.2000 | 72.0000 | 2.5378 | 3.0873 | 2.8125 |
| 4.2000 | 74.0000 | 2.5570 | 3.0923 | 2.8247 |
| 4.2000 | 76.0000 | 2.5762 | 3.0957 | 2.8360 |
| 4.2000 | 78.0000 | 2.5948 | 3.0989 | 2.8468 |

X DIRECTION RESULTS

CUBIC SITES

TEMPERATURE IS    4.20 KELVIN

FIELD TESLA

MAGNETISATION(BM)

0.00E 00                                                    0.26E 01

.................................................................

```
  8.000
     .
 10.000                          *
     .                              *
 12.000
     .                                *
 14.000
     .                              *
 16.000
     .                                *
 18.000
     .                                  *
 20.000
     .                                    *
 22.000
     .                                      *
 24.000
     .                                        *
 26.000
     .                                          *
 28.000
     .                                            *
 30.000
     .                                              *
 32.000
     .                                            *
 34.000
     .                                              *
 36.000
     .                                                *
 38.000
     .                                                  *
 40.000
     .                                                    *
 42.000
     .                                                      *
 44.000
     .                                                        *
 46.000
     .                                                          *
 48.000
     .                                                            *
 50.000
     .                                                              *
 52.000
     .                                                                *
 54.000
     .                                                                *
 56.000
     .                                                                *
```

X DIRECTION RESULTS

HEXAGONAL SITES

TEMPERATURE IS    4.20 KELVIN

FIELD TESLA

MAGNETISATION(BM)
0.00E 00                          0.31E 01

. . . . . . . . . . . . . . . . . . . . . . . . . . . . . . . . . . . . . . . . . . . . . .
```
                  *
0.000
         *
2.000
                   *
4.000
                *
6.000
                  *
8.000
                     *
10.000
                    *
12.000
                      *
14.000
                     *
16.000
                      *
18.000
                       *
20.000
                        *
22.000
                        *
24.000
                        *
26.000
                        *
28.000
                        *
30.000
                        *
32.000
                        *
34.000
                        *
36.000
                        *
38.000
                        *
40.000
                        *
42.000
                        *
44.000
                        *
46.000
                        *
48.000
```

X DIRECTION RESULTS

```
                    AVERAGE FROM BOTH SITES
                              TEMPERATURE IS    4.20 KELVIN
    FIELD TESLA
                                        MAGNETISATION(BM)
            0.00E 00                    0.28E 01
...............................................................
              *
     0.000
       .          *
     2.000
       .
     4.000              *
       .
     6.000                   *
       .
     8.000                  *
       .
    10.000                 *
       .
    12.000                   *
       .
    14.000                    *
       .
    16.000                     *
       .
    18.000                       *
       .
    20.000                        *
       .
    22.000                        *
       .
    24.000                          *
       .
    26.000                           *
       .
    28.000                           *
       .
    30.000                            *
       .
    32.000                             *
       .
    34.000                             *
       .
    36.000                              *
       .
    38.000                               *
       .
    40.000                               *
       .
    42.000                                *
       .
    44.000                                *
       .
    46.000                                 *
       .
    48.000
```

Z DIRECTION RESULTS

INPUT DATA FOR PRASEODYMIUM
LOWEST FIELD IS    0.00 TESLA
THERE ARE 40 STEPS OF    2.00 TESLA

APPLIED IN Z DIRECTION
MOLECULAR FIELD PARAMETERS USED=   -1.20     4.14
RESULTS ARE

| TEMP K | FIELD | MAG C | MAG H | AVERAGE |
|--------|-------|-------|-------|---------|
| 4.2000 | 0.0000 | 0.0000 | -0.0000 | 0.0000 |
| 4.2000 | 2.0000 | 0.1304 | 0.0005 | 0.0654 |
| 4.2000 | 4.0000 | 0.2398 | 0.0013 | 0.1206 |
| 4.2000 | 6.0000 | 0.3670 | 0.0020 | 0.1845 |
| 4.2000 | 8.0000 | 0.4743 | 0.0032 | 0.2387 |
| 4.2000 | 10.0000 | 0.5945 | 0.0042 | 0.2994 |
| 4.2000 | 12.0000 | 0.6956 | 0.0062 | 0.3509 |
| 4.2000 | 14.0000 | 0.8060 | 0.0081 | 0.4071 |
| 4.2000 | 16.0000 | 0.8994 | 0.0117 | 0.4556 |
| 4.2000 | 18.0000 | 0.9985 | 0.0152 | 0.5068 |
| 4.2000 | 20.0000 | 1.0914 | 0.0196 | 0.5555 |
| 4.2000 | 22.0000 | 1.1671 | 0.0293 | 0.5982 |
| 4.2000 | 24.0000 | 1.2485 | 0.0380 | 0.6433 |
| 4.2000 | 26.0000 | 1.3240 | 0.0496 | 0.6868 |
| 4.2000 | 28.0000 | 1.3907 | 0.0736 | 0.7322 |
| 4.2000 | 30.0000 | 1.4555 | 0.0988 | 0.7772 |
| 4.2000 | 32.0000 | 1.5154 | 0.1355 | 0.8254 |
| 4.2000 | 34.0000 | 1.5706 | 0.1907 | 0.8807 |
| 4.2000 | 36.0000 | 1.6360 | 0.3154 | 0.9757 |
| 4.2000 | 38.0000 | 1.6818 | 0.4683 | 1.0751 |
| 4.2000 | 40.0000 | 1.7556 | 0.6820 | 1.2188 |
| 4.2000 | 42.0000 | 1.8109 | 0.9649 | 1.3879 |
| 4.2000 | 44.0000 | 1.8628 | 1.2676 | 1.5652 |
| 4.2000 | 46.0000 | 1.9070 | 1.5664 | 1.7367 |
| 4.2000 | 48.0000 | 1.9440 | 1.8201 | 1.8821 |
| 4.2000 | 50.0000 | 1.9669 | 2.0170 | 1.9919 |
| 4.2000 | 52.0000 | 2.0003 | 2.1381 | 2.0692 |
| 4.2000 | 54.0000 | 2.0192 | 2.2203 | 2.1198 |
| 4.2000 | 56.0000 | 2.0369 | 2.2700 | 2.1535 |
| 4.2000 | 58.0000 | 2.0535 | 2.3003 | 2.1769 |
| 4.2000 | 60.0000 | 2.0689 | 2.3191 | 2.1940 |
| 4.2000 | 62.0000 | 2.0834 | 2.3312 | 2.2073 |
| 4.2000 | 64.0000 | 2.0970 | 2.3395 | 2.2182 |
| 4.2000 | 66.0000 | 2.1098 | 2.3454 | 2.2276 |
| 4.2000 | 68.0000 | 2.1257 | 2.3504 | 2.2380 |
| 4.2000 | 70.0000 | 2.1368 | 2.3538 | 2.2453 |
| 4.2000 | 72.0000 | 2.1472 | 2.3566 | 2.2519 |
| 4.2000 | 74.0000 | 2.1570 | 2.3591 | 2.2580 |
| 4.2000 | 76.0000 | 2.1663 | 2.3612 | 2.2638 |
| 4.2000 | 78.0000 | 2.1751 | 2.3632 | 2.2691 |

Z DIRECTION RESULTS

CUBIC SITES

TEMPERATURE IS    4.20 KELVIN

FIELD TESLA

MAGNETISATION(BM)

0.00E 00                              0.22E 01

...................................................................

```
 8.000
        .                         *
10.000
        .                            *
12.000
        .                              *
14.000
        .                                *
16.000
        .                                  *
18.000
        .                                    *
20.000
        .                                      *
22.000
        .                                        *
24.000
        .                                          *
26.000
        .                                            *
28.000
        .                                              *
30.000
        .                                                *
32.000
        .                                                  *
34.000
        .                                                    *
36.000
        .                                                      *
38.000
        .                                                        *
40.000
        .                                                          *
42.000
        .                                                            *
44.000
        .                                                              *
46.000
        .                                                                *
48.000
        .                                                                  *
50.000
        .                                                                    *
```

Z DIRECTION RESULTS

```
                        HEXAGONAL SITES
                              TEMPERATURE IS    4.20 KELVIN
    FIELD TESLA
                                     MAGNETISATION(BM)
                0.00E 00                0.24E 01
.....................................................................
      8.000
              *
     10.000
              *
     12.000
              *
     14.000
              *
     16.000
              *
     18.000
            .*
     20.000
            .*
     22.000
            .*
     24.000
            .*
     26.000
          .  *
     28.000
          .   *
     30.000
          .    *
     32.000
          .      *
     34.000
          .        *
     36.000
          .          *
     38.000
          .             *
     40.000
          .                *
     42.000
          .                   *
     44.000
          .                       *
     46.000
          .                          *
     48.000
          .                              *
     50.000
          .                                 *
     52.000
          .                                     *
```

Z DIRECTION RESULTS

AVERAGE FROM BOTH SITES

TEMPERATURE IS    4.20 KELVIN

FIELD TESLA

MAGNETISATION(BM)

0.00E 00                                        0.23E 01

```
........................................ ...........................
   8.000
     .                *
  10.000
     .              *
  12.000
     .               *
  14.000
     .                *
  16.000
     .                 *
  18.000
     .                   *
  20.000
     .                  *
  22.000
     .                    *
  24.000
     .                     *
  26.000
     .                    *
  28.000
     .                       *
  30.000
     .                       *
  32.000
     .                        *
  34.000
     .                          *
  36.000
     .                           *
  38.000
     .                              *
  40.000
     .                                *
  42.000
     .                                   *
  44.000
     .                                      *
  46.000
     .                                        *
  48.000
     .                                          *
  50.000
     .                                          *
  52.000
     .                                          *
  54.000
```